The DANCE *of* MOLECULES

The DANCE *of* MOLECULES

How Nanotechnology Is Changing Our Lives

Ted Sargent

Thunder's Mouth Press
New York

THE DANCE OF MOLECULES
How Nanotechnology Is Changing Our Lives

Published by
Thunder's Mouth Press
An Imprint of Avalon Publishing Group
245 West 17th Street, 11th Floor
New York, NY 10011

Library of Congress Cataloging-in-Publication Data is available.

ISB-10: 1-56025-895-0
ISBN-13: 978-1-56025-895-7

9 8 7 6 5 4 3 2 1

Printed in the United States
Distributed by Publishers Group West

To the memory of my grandparents:
generous, compassionate, inspiring, wise.

To my niece Clara Linder:
full of hope and promise, character and destiny.

Contents

Prologue

Imagine

The year Greta Garbo died of kidney failure in New York was the year I made up my mind to become a nanotechnologist. I was seventeen and Greta had me under her spell. Watching *Mata Hari,* I was beguiled, more devoted and unsuspecting than an ensnared Russian lieutenant. Mesmerized by the undulations of her hips, I wondered: Must I be content with just the dream of her? We knew so much about matter, more still about Greta, so why must it be absurd to imagine returning her to life? Not clone her, but construct her. Synthesize her, build her, and bring her to life from the atom up. Merge elements, transforming chemical into human form.

The necessary ingredients were at my disposal. Ample visual account was available describing Greta's detailed structure—her architecture and furnishings. Vestiges of silver oxide particles passed or masked light, retelling the projection of her shape. Her voice reverberated in my memory. Lionel Barrymore must have had the presence of mind to record the smell of her hair swooshing under his nose. Her blueprints must be available for download: I could upload my lips to her cheek.

Greta lives on in matter as well as in memory. Now, fifteen years after her death, her particles still surround us. Elemental

Garbonium wafts through the air we breathe. Back then, Greta's molecules were organized into a complex, elegantly coordinated superstructure. Proteins manufactured from the blueprint of her delicious DNA put muscle, fat, and liver in their splendid places. Oxygen and glucose fed hot red blood, giving Greta energy, life … enchantment.

No, Greta was not gone. It was just that her atoms were in all the wrong places, spread around the earth, but still somewhere in the material world. I wanted Greta back, integral. Her every sensory endowment was archived, and in the air were suspended the very elements which, appropriately arranged, had given her life. So why was I not dancing with Greta Garbo?

Living beings may be the most striking examples of the power of atoms, suitably organized, to yield a striking range of behavior, but they are not the only ones. Glass blocks bounce and bend light. Plastics flex while titanium bicycle frames gird. Hair coils and structural steel supports. Semiconductors conduct waves of electrons in harmonious concert. What if we could pick a property—bulletproofness, cancer-cell destructiveness—and specify and generate the molecules, and from these the materials, needed to implement our dream?

Some might seek to save lives using this astonishing power: create a device, fully integrated with the blood flow of the diabetic patient, to monitor and maintain blood sugar levels as in a healthy subject. Screen and filter cells continually for dangerous mutations. Remove pollutants from the air in our cities. Others might dream of taking lives: create new bullets so hard and streamlined that they could penetrate today's bulletproof vests. Build minuscule unmanned planes densely packed with explosives to bomb buildings. Engineer viruses that kill only the blue-eyed. Some might envision improving quality of life: create an active, hydrogen-powered muscle-suit that amplifies strength and finesse. Paint the walls of a room such that they display the real-time image of another place located far across the ocean, visually and aurally merging two homes separated by thousands of miles and obviating costly and polluting travel. Some might dream of improving their lives at others' expense: creating wireless tracking devices, microphones or cameras the size of dust, and puffing such imperceptible powders onto unsuspecting victims: e-stalking.

But enough fantasy—let's come back down to earth. Could we turn our dreams of tailoring matter to our needs into reality? It seems a tall order. We first need to measure: know which properties we experience and which atomic configurations a material possesses. Sense hair's bendiness and identify the atoms, molecules, and superstructures of which it is composed. Then we'd need to understand, trace out in full the relationships between structure and function. How does collagen's twist yield hair curly or straight? Next we'd need to invert the problem: place an order for hair not curly, but veering 90° to the southwest after each inch-long straightaway, and then specify the molecules to make it work its magic. Finally, we'd need to manufacture our molecules and induce them to curve as needed. They'd have to do the same thing every time, or, better yet, we'd need to architect them such that the occasional modest slip-up didn't ruin the overall effect. Here biological systems would inspire us at every turn, their every instance of an organism, each leaf and snail, not imperfect, only unique.

But today we analyze better than we synthesize, and that's the crux of the problem. Scientists deconstruct matter into its elemental constituents, but we are not yet able to trace out fully the links between the molecular—the nanoscopic—and the macroscopic realities we all encounter day by day. We can know the details of chemical composition intimately yet still not fully grasp how function arises from structure. Sequencing the human genome has not pinpointed the Garbo Allure gene. Height, disposition, and predilection are instead sprawled across abstruse molecules of DNA. For now, Garbonics remains a humanity, not a science.

Today we can marvel at Nature's glorious creations, but when it comes to designing our own using Nature's Lego blocks, we are all thumbs. Had we the benefit of an atom-by-atom plan of Greta's lush three-dimensional architecture, resolved with nanometer scale bar, we would still be unable to erect her. Both conceptually and chemically, we can neither fully explain nor build the links between the bottom—the arrangement of atoms that constitute matter—and the top—complexity, subtlety, function, and dysfunction. As we experience it today, macroscopic reality wends its way mysteriously out of nanoscopic form.

This is where nanotechnology comes in. Nanotechnologists have as their goal to design and build matter to order, specified by a functional requirement. Nanotechnology is coordinated movement, a choreographed dance among atoms and molecules to achieve a desired effect. It harmonizes within Nature's own set of rules to coax matter to assemble into new forms. The resulting materials exhibit striking beauty when viewed in an electron or optical microscope, often even with the naked eye. Their purpose: to produce breakthroughs in medicine, energy, and information.

Nanotechnology is not a new science. For four billion years, Nature has organized atoms into simple molecules; molecules into proteins; proteins and sugars and fats into complex societies of cells; and cells into the life that surrounds us. Nature builds using a finite array of atomic elements, the periodic table. She is rigorously disciplined, limiting herself to a small set of simple, but powerful, rules. Physics is simple but subtle, quantum mechanics pioneer Paul Ehrenfest used to say. With a modest set of elements deployed subject to rigorous rules, Nature invents limitless variety, beauty, form, and purpose.

For centuries scientists have exploited Nature's ready-made molecular assembly line. We have linked molecules to form long, perfect polymer chains with predictable properties, creating Tupperware and rubber tires. We have introduced lumps of imperfect, impure material into a vacuum chamber, let atoms evaporate, and grow from them strikingly perfect crystals of defined shape, size, and orientation. We have controlled how these designer materials produce light, conduct electricity, and respond to the touch.

Nanotechnology is an intersection—a confluence at the heart of contemporary science. It is where the latest breakthroughs in chemistry, physics, and biology merge, mix with engineering and medicine, and produce chips, diagnoses, and therapies that no sequestered specialist would generate. Nanotechnology produces convergent thinking when representatives of various mind-sets meet, learn one another's languages, and gather the ideas that result when paradigms collide.

Specialization evolved as a necessary response to science's rapid rise in the Renaissance. The tree of knowledge diverged into separate branches: chemistry, physics, biology. In the last century, each branch spread into twigs—biochemistry into pharmacology, drug discovery, pharmacotoxicity. Now the culture of research in the scientific, engineering, and medical communities is undergoing a second renaissance. Researchers out in the twigs are recognizing that they all are connected to—nourished by—the same trunk. Mechanical engineers are using nanometer-sized probes to pull on biochemists' proteins, studying their remarkable tensile properties. Electrical engineers are working with biologists to grow not just circuits, but cells, on a silicon chip. Information theorists are learning more about their field by observing with awe the native capacity of DNA to correct errors during transcription. Fundamental scientists, once focused narrowly on the quest for understanding in its purest form, are taking pride in changing people's lives through their research. Where once this rendered science impure, now it makes it personally as well as profoundly relevant. For their part, engineering and medical researchers are not waiting for scientific breakthroughs to mature into completely understood, fully controlled technologies before beginning to use them to create new therapies, diagnostic tools, and communications devices.

If we had a comprehensive understanding of how a given molecular form results in observable function, could we then construct materials, perfect down to the placement of atoms, which fit our structural requirements? In their techniques of building matter to a specification, nanotechnologists were once divided into two camps, the top-downs and the bottom-ups. "Top" lies at the summit of the hierarchy of function: the useful purpose endowed by a macroscopic property, the desired goal. "Bottom" refers to the smallest material size scale imaginable, the realm of atoms and molecules. Top-down and bottom-up nanotechnologists are both matchmakers for atoms and molecules, but they use different tactics.

Top-down nanotechnologists emerged from electrical engineering, the field that gave us the astonishing success of microelectronics in the second half of the twentieth century. Well before we accessed the nanometer, electrical engineers were systematically working their way down from the goal of producing a versatile, efficient, fast-computational engine, to specifying and manipulating matter, forming shapes of metal and semiconductors that, once connected, constitute the billion-transistor integrated circuits inside our laptops. As of the 1960s, computer chips have been built using lithography, screen-printing that has allowed us to imprint forms as small as 100 nanometers. One-hundredth of a cell. One-thousandth of a human hair. One-ten-millionth the diameter of an elusive celluloid seductress like Greta Garbo.

From this tradition of top-down premeditated arrangement of matter emerged an early approach to nanotechnology. Fear catalyzed the process: would the powerful economic engine of information technology grind to a halt when engineers could not build circuits smaller than conventional lithography could imprint? Breakthroughs were needed to control morsels of matter smaller than 100 nanometers—though still much bigger than individual atoms and molecules. Engineers needed to develop the ability to craft matter on the nanoscale.

Long before the integrated circuit revolution, physicist Richard Feynman, the surfing, womanizing winner of the 1965 Nobel Prize in Physics, articulated what became the ultimate dream of the top-down camp. He gave his name to a family of—appropriately—sperm-like diagrams that describe fundamental interactions among elementary particles. He addressed the 1959 annual meeting of the American Physical Society in a talk titled "There's Plenty of Room at the Bottom." Feynman proved from the laws of physics and the known properties of matter that the entire twenty-four volumes of the *Encyclopaedia Britannica* could feasibly be written on the head of a pin. Four decades after Feynman's talk, Gerd Binnig and Heinrich Rohrer of the IBM Zürich Research Laboratory wrote the word *IBM,* atom by atom, using the scanning tunneling microscope for which they had won the 1986 Nobel Prize in Physics. Had they kept writing, they could comfortably have fit the encyclopedia onto the tip—not the head—of their pin. Macropaedia, Micropaedia? Nanopaedia.

But does meticulous, self-conscious atom-arranging bring us closer to Greta Garbo? To push around carbon, oxygen, and nitrogen atoms to form 50 kilograms of luxurious closet-Swede is a major enterprise. The math tells the story:

{Mass of Greta Garbo ~ 50 kg} ÷ {Mass per carbon atom ~ $2x10^{-23}$ grams} = $2x10^{27}$ atoms to be arranged

Assemble a billion billion billion atoms? Even a team of twenty graduate students working full throttle—Igor and Sergei will each take a leg if Tung-Wah handles hands—could not hope to execute the project in the five years of a doctoral dissertation. That it would take longer than I am willing to wait is only a fraction of the problem. Would you trust an engineer to design Greta Garbo? When top-down atom-pushers design devices, circuits, and systems, they insist on explicit nanometer-scale control over what they will build. Intel engineers in ultra-clean fabrication facilities dress in canary-yellow spacesuits to prevent a single particle from alighting on your computer chip. There is centralized planning of every nanoacre. If one transistor goes off the grid, the computational engine grinds to a halt: the authorities insist upon rigid perfection.

In contradistinction, Nature builds imperfect things that work perfectly. Each maple leaf is atomically unique, but incontrovertibly a member of its class. Nature has no flaws, only miracles. Thus inspired, the new generation of nanotechnologists seeks to learn from and exploit self-organization of matter. Nobel laureate Jean-Marie Lehn spoke of the sociology of molecules: once we know how each molecule behaves, and how it interacts with its fellow citizens, we can predict the communities these molecules will form. From knowledge of the rules each molecule obeys, we can predict the emergence of the structure and function of a material—a molecular society and, ultimately, a material culture. Beating heart muscle. Pollutant-filtering molecular sponges. Energy-harvesting solar cells.

Let us embark upon a journey into the world of nanotechnology. Let us see how far we have come in persuading Nature to fashion matter after our needs, in using refined control over atoms, electrons, and photons to better human existence. Let us examine how the latest breakthroughs are revolutionizing human health, environment, and information. Humbled before Nature's achievements, let us inquire as to our own limitations, and contemplate what responsibilities arise in the face of our newfound abilities.

The DANCE *of* MOLECULES

Introduction

Discover

Inspired, guided, and aided by Nature's biology, nanotechnologists turn the physical into the functional, chemistry their implement. In the physical realm, the laws Sir Isaac Newton knew many centuries ago serve us well today. Trains on tracks rolling, doors on hinges opening, clocks' pendula swinging. Newton described the physics of the worldly and the tangible, our day-to-day experience. He gave us insight into light, showing systematically how to direct images using lenses. Those scientists who, in the 18th and 19th centuries, stood on Newton's shoulders saw farther, understanding and then controlling the flow of electricity. By the late 1800s, James Clerk Maxwell had unified the electric and magnetic forces and, using the power of mathematics, had derived how waves of light ripple through space and time.

Questions remained nevertheless. What made up this electric current that flowed in circuits? Was it built out of particles like Democritus' indivisible atoms? The same question could be asked of light: if it was a wave, then surely it could be decomposed into dimmer and dimmer oscillations—or with light too was there a limit to reductionism?

Discovering the Grammar of the Nanoworld

The answers to these questions have come in the past 100 years. Albert Einstein proposed in 1905 a quantum theory of light to explain a series of experiments whose sense had eluded others. Simple measurements of the flow of electricity from a metal illuminated by the colors from a lamp forced Einstein to posit that light came in indivisible packets known as photons. In 1927, Werner Heisenberg, who was among the pioneers of quantum mechanics, claimed that the more precisely the position of a particle such as an electron is determined, the less precisely its momentum, or speed, may be known. In classical physics, it is a matter of faith that you can know simultaneously how fast a car is going and where it is located. In quantum physics, when nanometer particles are at play, it is a matter of fallacy.

We must delve into the chapter in Nature's rulebook titled Quantum Mechanics to understand what will result from self-assembly and self-organization among basic particles. We discover then how Nature authors her infinitely variegated masterpieces using an alphabet of atoms linked via an exceptionless grammar.

Much of quantum mechanics derives from one simple but surprising assumption: electrons are not only discrete particles of charge, but also waves. Light, long known to be a wave, is at the same time made up of particles: photons, quantized bundles of optical energy. The term wave-particle duality describes this paradoxical picture, electrons and photons as countable packets of wave.

Quantum mechanics gives nanotechnologists a set of rules to exploit, and nanotechnologists have already taken advantage of its possibilities, one striking example being the quantum size effect. Just as the oscillatory frequency of a guitar string is determined by its length—the pitch governed by the position of finger on fret—so too does the length to which electron waves are confined determine the set of frequencies at which they can oscillate. As with the defined overtones of a guitar string, the energies that result form a ladder of resonances. Long strings produce rungs so closely spaced as to be nearly indistinguishable, and the quantum world fuses into continuous classical physics. This illustrates the Correspondence Principle, the link between Heisenberg's and Newton's worlds at their gray boundary. For short strings—tightly confined electrons—the rungs of the energy ladder are well separated and clearly discrete, and one cannot miss the quantum mechanical nature of the electrons.

Quantum mechanics provides a means of tailoring matter, for changing a particle's size will change the energies electrons may adopt. These energies determine the colors of light that a material will absorb and emit. The quantum size effect turns monochromatic matter into Technicolor designer materials. Bill bissett, the Canadian poet, wrote in "th tomatoe conspiracy aint worth a whol pome:"

Very few peopul
    realize ths but altho yu have 5 or 6
billyun peopul walkin around beleeving

that tomatoes ar red they ar
        actually blu an ar sprayed
red to make ther apperance
    consistent with peopuls beleef

Through the quantum size effect we can customize materials beyond our previous expectations. To find size-tunable blue bananas, turn right at the glorious bouquets of ultraviolets. If you reach the bin of infrared Delicious apples, you've gone too far.

Mastering the Nanometer Alphabet

Nanometer grammar may seem exotic, but today we understand it well. Nanotechnologists piece together letters, words, sentences, and paragraphs that have meaning. Our ultimate triumph will be to construct narratives as intricate as life, erected from an atomic alphabet.

We have at our disposal slightly more than 100 distinct atoms, each one built from electrons orbiting a nucleus. Some electrons carve out tight trajectories, hugging the center, while others trace sweeping arcs. The outer electrons interact most strongly when atoms approach one another. Atoms thus concern themselves little with the details of their inner cores: the fate of relationships among atoms—chemistry—is written in their outermost electrons.

A nanotechnologist will ignore the details of atoms' inner lives, instead making casual reference to mass, charge, and valence. Such objectification gives rise to stereotype, prejudgment: "That sulfur, she ain't never met a gold atom she didn't bond with." Though nuclei are intricately detailed, to a chemist they are but labels predicting character, betraying known propensities: tropes, archetypes.

Composing Words: Building Molecules

Nature and Chemist have for centuries joined atoms to compose materials with properties strikingly different from their constituents. Sodium is a metal that reacts violently with air. Chlorine is a lethal gas. Sodium chloride, their fusion in equal ratio, is table salt, a stable white crystal necessary to life. Suddenly, by collaborating, the members of our modest alphabet gain versatility, potency. The number of possible combinations of elements in the periodic table is vast, the diversity of powers and properties lush. They are as varied as the world that surrounds us, filled with solids, liquids, gases; granite, graphite, diamond; hibiscus, wildebeest, Tasmanian winter truffle. Unbelievable though it might seem, Willamette pinot noir and Barossa shiraz are built using the same alphabet.

The number of possible combinations of elements in the periodic table is vast, the diversity of powers and properties lush.

Might the set of combinations possible, and therefore the properties available, be even more vast and varied than what we and Nature have fashioned thus far? One single alphabet existed before Shakespeare as well as after; yet Shakespeare left in his wake a voluptuous, multihued, raucous English garden—one grown from nothing but mundane letters. We have yet to use up our power to express anew using twenty-six letters, and our atomic alphabet similarly shows no signs of creative exhaustion. This is the nanotechnologist's premise: that there remain to be written enthralling material oeuvres not yet authored by Nature.

This might not be true if we could only combine two elements at a time—sodium and chlorine for salt, iron and carbon for steel, two oxygen atoms for the oxygen molecule in air. "To do is to be": the range of meaningful sentences we can construct using two-letter words is limited. Molecules, on the other hand, can be built from any number of atoms. Two in salt; three in water; four in hydrogen peroxide. The proteins in our bodies are made of millions of atoms, as is DNA, molecule of heredity. Molecules thus vary tremendously in size, much more so than English-language words. From a nanometer to hundreds of nanometers: *Ja* to *Donaudampfschiffahrtselektrizitaetenhauptbetriebswerkbau-unterbeamtengessellschaft*.

Chemists know that the inclinations of some atoms and molecules to bind, and others to repudiate one another, are not for us to decide. These are Nature's rules, awaiting our fuller discovery and inviting our invention of new harmonies within their established frameworks. It is as with Bach's counterpoint: neither did he invent, nor did he define, our notions of assonance and dissonance. The composer uncovered strategies within a highly constrained framework that provided aural satisfaction, and he proved that, with a modest alphabet, restrictive rules, and boundless creativity, his capacity to excite new strains of emotional response in his listeners was without limit.

If only fated atomic unions can be consummated, then we can do nothing more than create opportunities: make an introduction, raise the temperature, ensure circulation of yin and yang within a purple velvet piano lounge. We can but catalyze fate, not alter destiny. There will be no nanomachines forcibly assembled atom by atom. Instead, to inform their craft, chemists have long studied the likes and dislikes of atoms and molecules, brought about fated pairings en masse, and harvested their progeny. Chemists do have the power to bring together well-matched partners that otherwise might not have found one another: merging lamb and mint, introducing duck to kumquat.

Making Up Words: Nanotechnology's Early Lexicon

Early nanotechnologists discovered some intriguing new pairings. Remarkable molecules discovered in the 1980s ushered nanotechnology into the lexicon and consciousness of scientists and engineers. The new nanoparticles were built from the most mundane of atoms: carbon, which insinuates itself into nine out of ten chemical substances we know. Without it, no pencils, no diamonds would exist—neither accounting nor materialism. Carbon is necessary to life: without it we would have no DNA, proteins, sugars, or fats.

Two forms of pure, ordered carbon have long been well known to us: diamond and graphite. The strength of each material derives from its atoms' arrangements and interactions. The atoms in diamond are strongly bonded together in three dimensions, conferring hardness from any angle. Graphite's atomic sheets are individually strong, but slide against one another, making the material soft as pencil lead. The contrast between these properties—held by two different materials made of identical atoms, just differently arrayed—illustrates the critical importance of atomic arrangement.

In the mid-1980s, the hitherto staid story of the carbon atom's regular arrangements took an exciting twist. Harold Kroto at the University of Sussex was interested in the chemistry inside a class of stars, the red giants that are cooler than our sun and typically 10 to 100 times bigger. From the way in which the stars absorbed light Kroto believed that they might contain many carbon-rich molecules of particular interest to him; he wanted to know how these stars were formed. Kroto searched for a way to reproduce in the laboratory conditions akin to those near red giant carbon stars and thereby produce his molecules on Earth. This yearning led him to contact Richard E. Smalley. At Rice University in Texas, Smalley routinely produced extreme conditions inside his chemical chamber, generating clusters—aggregates of atoms bigger than a typical molecule. Smalley used a laser to vaporize material inside his chamber into a reactive gas of atoms and then studied the products of his reaction.

In 1985, Kroto, Smalley, and Robert Curl came together to use Smalley's instrument to study how to form clusters of carbon. Kroto hoped to see clues to what was being formed in the hot parts of stars' atmospheres. Graduate students Jim Heath, Sean O'Brien, and Yuan Liu vaporized carbon, which condensed in their chamber. The vaporized atoms formed carbon clusters of a range of sizes, varying from a few atoms to many hundreds. These clusters could then be analyzed, first for size.

Certain specific sizes—"magic numbers" of atoms per molecule—predominated. The most striking was C60, made from sixty carbon atoms. The team found conditions in their reaction chamber—pressures, gas flows, vaporizing laser pulses—that led consistently to C60 and also to the almost-as-magical C70. Clearly these molecules were meant to be: polygamous atomic marriages made in heaven. Noting how stable the fated molecules were, the researchers suspected that they were seeing not the long chains Kroto had been hunting for, but something different: a cage, a closed shape, self-contained.

Clearly these molecules were meant to be: polygamous atomic marriages made in heaven.

Kroto thought of American architect Buckminster Fuller's geodesic dome, which he had first seen at the Montreal World's Fair of 1967.

Kroto wondered whether the dome might have exactly sixty vertices. According to Kroto, "That night Jim and his wife experimented with cage models and Rick with paper hexagons.... Rick tried them and found that the flat hexagon-only model curled up, and on adding 12 pentagons a closed ball with 60 vertices grew. When he tossed the ball on the table the next morning we were ecstatic. It was so beautiful it just had to be right." The team gave the structure the fanciful name buckminsterfullerene. After eleven days of experiments and writing, they sent a manuscript titled "C60: Buckminsterfullerene" to *Nature*. The opening few sentences of the paper, published in November 1985, tell succinctly the story of their new molecule:

> During experiments aimed at understanding the mechanisms by which long-chain carbon molecules are formed in interstellar space and circumstellar shells, graphite has been vaporized by laser irradiation, producing a remarkably stable cluster consisting of 60 carbon atoms. Concerning the question of what kind of 60-carbon atom structure might give rise to a superstable species, we suggest a truncated icosahedron, a polygon with 60 vertices and 32 faces, 12 of which are pentagonal and 20 hexagonal. This object is commonly encountered as the football shown in Fig. 1. The C60 molecule which results when a carbon atom is placed at each vertex of this structure has all valences satisfied by two single bonds and one double bond, has many resonance structures, and appears to be aromatic.

Curl, Kroto, and Smalley went on to win the Nobel Prize in Chemistry in 1996 for their discovery. The C60 structure and its properties have formed a vast source of research opportunities.

The team continued their investigations, trying to react buckyballs with other compounds. They failed, giving further evidence that C60 was stable, a closed shell. The group even managed to shrink-wrap their buckyballs around metal atoms. The atom inside, they used a laser beam to tighten the carbon cages by two carbon atoms at a time. When the metal atom, finite in size, resisted further confinement, shrinkage ceased. The shell had collapsed just enough to fit exactly around the metal atom, and the researchers had learned more about the size of the cage.

By the early 1990s, another method of vaporizing the carbon atoms in graphite allowed scientists to produce vast quantities of buckyballs. And as it turns out, Nature had been making buckyballs for millennia; we simply hadn't analyzed them. In a burning candle, hot flame turns wax to vapor, molecules that contain a mixture of carbon, hydrogen, and oxygen. Amidst the hot-burning soot produced in the yellow part of the flame are buckyballs. They exist also in interstellar dust of interest to Kroto, and in geological formations on Earth.

With their successful production in the abundant quantities needed for manufacturing, the properties of C60 have been explored in detail and many have been usefully applied. Chapter 4, Energize, explores how buckyballs increase the energy conversion efficiency of cheap, flexible, wearable solar cells. Chapter 5, Protect, discusses how certain buckyballs show promise for light-activated cancer therapy, but also raise questions as to their effect on the environment. In Chapter 9, Convey, buckyballs are shown to have resulted in materials that allow one beam of laser light to control another, all within trillionths of a second.

Pseudo-spheres of sixty carbon atoms are not the only intriguing nano-objects to have been discovered in recent decades. Nanotubes, or buckytubes, are also produced by vaporizing graphite in a controlled environment. Like buckyballs, they have nanometer diameters; but unlike buckyballs, they can be hundreds of nanometers or longer along one dimension. Carbon nanotubes were discovered in 1991 by Sumio Iijima at the NEC laboratory in Tsukuba, Japan. Iijima was studying the production of buckyballs made by passing current through graphite. He inspected the results of his reactions by looking at the nanotubes using a transmission electron microscope with nanometer resolution and he saw long, narrow tubes. Soon Thomas Ebbesen and Pulickel Ajayan, researchers with Iijima, showed that they could produce more abundant quantities of nanotubes by varying conditions in the production chamber. Carbon nanotubes are rolled graphene sheets—the two-dimensional arrays of carbon atoms that usually stack to form planar graphite—into baklava-like cylinders.

One-dimensional nanometer-diameter objects are of great interest to theorists and experimentalists alike; they were expected to have unusual strength and new electronic properties. Built from varying numbers of layers, though, these nanotubes would not lend themselves to ready understanding. Instead, the behavior of the collection was dominated by tube-to-tube variations: many individually exquisite melodies, juxtaposed unsynchronized, make a cacophony.

In 1993, nanotubes that were one carbon sheet thick were synthesized for the first time, both by Donald Bethune and colleagues at IBM's Almaden Research Center in California, and also by Iijima's group. Adding metals such as cobalt to the graphite resulted in a fine tube with single-layer walls. These single devices proved much simpler to understand than the original nanotubes, their properties purer.

Nanotubes, it turned out, vary not only according to their size; even those of the same diameter can exhibit a range of properties.

There is more than one way to wrap a bottle of wine: line up the edge of the sheet of wrapping paper with the height of the bottle, or orient the bottle and paper at 30°, or for that matter at any other angle. The choice isn't quite so free with nanotubes, for they need to link up atom by atom after they've been curled to form a cylinder. But there remain many ways to make a perfect join, rolling them at different angles relative to the ordered rows of atoms. The rolling angle determines the properties of the nanotube. Some tubes behave like semiconductors, others like metals. Metals conduct intrinsically: they supply not only the path along which electrons can flow, but also the electrons necessary to conduction. Semiconductors provide the tracks, but not the electrons that actually convey electrical current. They need a source of electrons: light, heat, an electric voltage, or philanthropic electron-rich atoms can donate to the cause of conduction.

With this understanding of the nanotubes at their disposal, researchers explored the relationship between structure and behavior. The cylinders are so small that they force electrons to take on an unusual tubular shape, mimicking the structure of the buckytubes themselves. Looking outward radially from the center of the tube, an electron is confined by the graphene sheet one carbon atom thick. Traversing the tube circumferentially, an electron head must join with its own tail. And these tubular electrons can flow in only one direction, the longitudinal one. The energies with which the electrons flow come from their detailed configuration radially and circumferentially about the tube.

Light, heat, an electric voltage, or philanthropic electron-rich atoms can donate to the cause of conduction.

Since the electronic properties depend so strongly on the details of nanometer diameter and rolling and end-fastening, one learns little by looking at a collection of different nanotubes. In 1994, Charles Olk and Joseph Heremans at the General Motors Research Laboratory managed instead to measure properties of individual tubes. They showed that, as theorized, some nanotubes are metallic, others semiconducting. The understanding of the relationship between structure and function has enabled recent devices using the tubes: single-wall nanotubes have been used to make transistors, the building blocks of computer chips, based on one molecule (see Chapter 7, Compute).

Buckyballs and carbon nanotubes illustrate both the promise and challenges of much early engineering at the nanoscale. They are "bespoke" molecules, which Nature tailors to our needs. Since Nature is so versatile with her bag of tricks, the researcher seeks ways to make the desired size and color of bunny pop out of her hat, and to do so consistently. Smalley and colleagues, for example, found the conditions of laser pulse intensity and duration such that buckyballs were formed preferentially over all other possible carbon clusters. In bottom-up nanotechnology, Nature does the work, and we harvest the fruits of her labors. If Nature's works are beautiful but abundantly variegated, how do we direct her to build what we need, or alternatively, how do we sort the wheat from the chaff? The need to manage, indeed to exploit, rather than be limited by, heterogeneity in Nature's prodigious productivity is one theme explored in this book.

Another early nanoparticle permeates this book: quantum dots, crystals a few nanometers in diameter. The choice of elements combined to make these crystals determines their baseline properties.

Further tuning comes through the quantum size effect. A semiconductor found in the wild must, by definition, have a bandgap—an energetic no-electron's land. The size of this gap governs which colors of light are absorbed. Light incident on a semiconductor will not, if its photons' energy is smaller than the bandgap, manage to elevate the electrons within to higher energies. They may try to boost them to the next level, but if their energy is insufficient, the electrons will simply not reach the second story of the building. And researchers have recently shown that they can master quantum dot shape as well, with breathtaking images coming out of the lab of Paul Alivisatos of University of California at Berkeley. In 2004, they built shapes such as tetrapods whose arms were made out of a controlled sequence of materials.

Quantum dots and their brethren have seen wide application in the early days of nanotechnology. They play starring roles in this book in diagnosing cancer at the earliest malignancy (Chapter 1, Diagnose), connecting people and information technology in new ways (Chapter 8, Interact), building an optical Internet nimbly linked to our growing wireless communications infrastructure (Chapter 9, Convey), harnessing the sun's power in abundance and with heightened efficiency (Chapter 4, Energize), and mimicking biology's prowess in building molecules rigid and bendy, green and pink, sweet and savory (Chapter 6, Emulate).

Stringing Together Sentences

Nanoparticles are, within our analogy of constructing meaning from the letters of a finite alphabet, rather long, elegant words. They are neologisms in the nanoscale vernacular, additions to scientists' active molecular vocabulary. On their own, though, they still leave us far from talking in complete sentences. And without syntax, the rules connecting words, we are in a world of representation but not ideas.

Jean-Marie Lehn, co-winner of the 1987 Nobel Prize in Chemistry, defined his field of supramolecular chemistry as follows: "Atoms are letters. Molecules are the words. Supramolecular entities are the sentences and the chapters." Lehn advanced our understanding of how molecules interact with one another, focusing on the idea of recognition among molecules prevalent in biology. Without molecular recognition we would be dead: the communication of signals along our systems of nerves relies on it. For our nervous system to work, the nerve cells must be able to tell the difference between the almost-identical atoms, spheres that have the same charge; their only difference is a slight one in size. Proteins in our cell membranes boldly distinguish atoms with sub-nanometer subtlety. Our own immune system provides another elegant example. An antigen is anything that can trigger an immune response: bacteria, viruses, or pollen. An antibody is a molecule produced by our immune system to identify and then neutralize foreign objects. In our immune system, each antibody is responsible for recognizing a specific antigen. The intricacies of molecules' shape and chemistry provide for lock-and-key fits—Cinderella's feet sliding perfectly, and exclusively, into her Manolo Blahniks.

Lehn's passion for molecules' interactions—the sociology of molecules—comes from the recognition that they are rarely alone, and that in the company of other molecules their relevant properties often emerge. Water, as a molecule, is simple: one oxygen, two hydrogens—a triangle. One can study individual water molecules, and indeed many scientists do, but the individual molecule admits no notion of melting, evaporation, freezing. The collection has totally different properties from the individual. Lehn sought not to reduce all materials to their molecular and atomic constituents, but instead to deduce properties from underlying nanostructure. One cannot reduce the melting or freezing of water to a single molecule, for in the context of a single molecule the idea is without meaning. But from knowledge of the properties of these molecules germane to their mutual interactions, one can deduce collective properties.

Lehn thus set about building molecular entities that could tell small from large spheres—molecular recognition at the smallest scale, the length of the atom. Charles Pederson, Lehn's co-laureate, had shown in 1967 that he could devise molecules whose shape and size resulted in binding to different atoms. It was predominantly shape—the fit between his molecules and the ions of interest—that led to this selectivity. Their co-worker Donald Cram then showed that he could design host molecules that would bind their guest atoms strongly and selectively: one of these could bind sodium ions half a million times more strongly than lithium ions. These atoms had similar charge and general shape, differing substantively only in size. Cram and Lehn took on the challenge of showing a diversity of specific binding designs for ions of all shapes, sizes, and charges. They rapidly showed practical utility as well as fundamental interest. They extracted toxic atoms from the environment without affecting the other atoms. Cram also showed that he was able to sort

> **Jean-Marie Lehn sought not to *reduce* all materials to their molecular and atomic constituents, but instead to *deduce* properties from underlying nanostructure.**

small proteins that differed only in that they were mirror images of others; his molecules could tell left from right. Cram and Lehn managed to build new molecules that mimicked biological enzymes.

Lehn, in constructing molecular sentences, worked not only with nouns—individual molecules—but with verbs as well. His focus on pairing in the service of action leads from a descriptive nanoscience into molecular storytelling. Lehn's molecules did not constitute the first nanoscale narratives written by human hand, for chemists had been composing new reactions for centuries. But the ability to design new subject–object interactions at will has enabled a considerable thickening of the plots we can weave.

Learning to Read

How are we so sure that we know all of this? It is often asked of nanotechnologists how they can be so confident that they are succeeding in engineering new properties from new structures when the structures in question are 1,000 times smaller than what we can see with the best optical microscope. Fortunately, we *can* see objects this small—but not with a conventional optical microscope, which does us no good at the nanometer lengthscale. In all microscopes there is a limit of performance—a smallest object that can be resolved. The limit is determined by the wave nature of the ray being used to trace out the two-dimensional image of the object. Conventional optical microscopes, so critical to our understanding of biology since the 1600s, were improved until they reached a limit determined by light's wavelength. Users cannot distinguish features smaller than a few hundred nanometers.

The use of another wave-particle—a much smaller one than light—was thus mandated. The 1986 Nobel Prize in Physics went in part to Ernst Ruska of the Max Planck Institute in Berlin for "his fundamental work in electron optics and for the design of the first electron microscope." In his early twenties, as a student at the Technical University of Berlin in the late 1920s, Ruska explored how a magnetic coil—turns of a wire through which current flowed—could direct a beam of electrons. He discovered how to focus his beam to a point, making a lens. Using electrons he irradiated the object and recorded the image on a fluorescent screen or a photographic plate. With the aid of two or more lenses Ruska was able to increase the magnification. By 1933 Ruska had built a microscope based not on photons, but on electrons, one that produced an image strikingly superior to that of an optical microscope. And by the late 1930s Ruska had worked with colleagues at Siemens to create the first commercially available, mass-produced electron microscope.

Images taken using a modern-day transmission electron microscope (TEM) allow resolution better than 0.1 nanometers, comfortably resolving individual columns of atoms in crystals. In recent years, transmission electron microscopes have gained a new capacity: the ability not only to locate physical structure, but also to determine chemical composition. Different atoms interact distinctively with electrons of varying energies, ensuring that TEMs that resolve the energy of transmitted and scattered electrons contain signatures of the elements making up the materials under study. Now we can image structure and fingerprint composition at the atomic scale.

A new technique of seeing, and indeed manipulating, atoms emerged more recently. Scanning tunneling microscopes (STMs) were born in what became the formative years of nanotechnology, the early 1980s, during which the field began to gain identity. The other half of the 1986 Nobel Prize in Physics went to Gerd Binnig and Heinrich Rohrer of IBM Research in Zürich for "their design of the scanning tunneling microscope." Binnig and Rohrer's is by no means a conventional microscope. It does not project an image of an object, transmitted or reflected, onto a screen. Instead it scans the surface of a sample as if reading Braille, tracing its finger systematically back and forth over the bumps made by atoms. The stylus—as sharp as one atom—senses the amplitude variations in the surface of the sample and reports them back to a computer, which displays a reconstructed image.

The microscope took ingenious engineering on the part of Binnig and Rohrer. They were looking for movements of their stylus of a scale much finer than the diameter of an atom, so any vibrations in their surroundings would ruin the image; their mechanical design had to have tremendous finesse. Instead of building their microscope on a structure that could vibrate, they isolated it from the outside world, placing it on a heavy magnet floating freely in a dish of superconducting lead. Less imposing but still effective methods of vibration isolation have since been developed.

Binnig and Rohrer's is by no means a conventional microscope. It scans the surface of a sample as if reading Braille.

Another insight was not to run their stylus at a constant elevation over the surface, but instead to run it at constant "feel." To sweep a fragile probe, whose tip was a single atom, along the surface until it crashed into a mountain would ruin the probe. Instead, Binnig and Rohrer used the tunneling current—the weak but telling flow of electrons from tip into sample—as a proxy for the distance between the atom at the tip of the probe and the surface of the sample whose topography it was exploring. Electrons, as quantum mechanical waves, are by the uncertainty principle not localized exclusively to the tip, but exist with low but finite probability in the atoms of the sample to which the tip is brought very close. A tunneling current flows, with electrons flying through forbidden space and into the sample. Since this small current decreases rapidly with separation between sample and tip, Binnig and Rohrer had hit upon an ultra-sensitive surrogate for spacing. By controlling this current, they kept their tip at a fixed distance from the surface of the sample. By monitoring the position of the tip that achieved this constant current, they measured the topography of their sample.

The stylus was scanned mechanically in the sample plane with a horizontal resolution of 0.2 nanometers. The vertical control over the tip, used to keep constant the local tunneling current, was 0.01 nanometers. Individual atoms looked like Everest. The images produced using scanning tunneling microscopes over the past two decades have been spectacular. One from the group of Don Eigler at IBM is shown on page 21.

What is most significant to nanotechnology today about these methods of microscopy is not so much their novelty, but the fact that they have become routine. Nanotechnologists image their samples as a matter of course. Not to do so would be to attribute function to structure without proving the link, and thus would skirt the central premise of nanotechnology, the building up of macroscopic utility from nanoscopic form.

New methods of analysis at the atomic and molecular scale have not been limited in their impact to the scientific; their legacy is

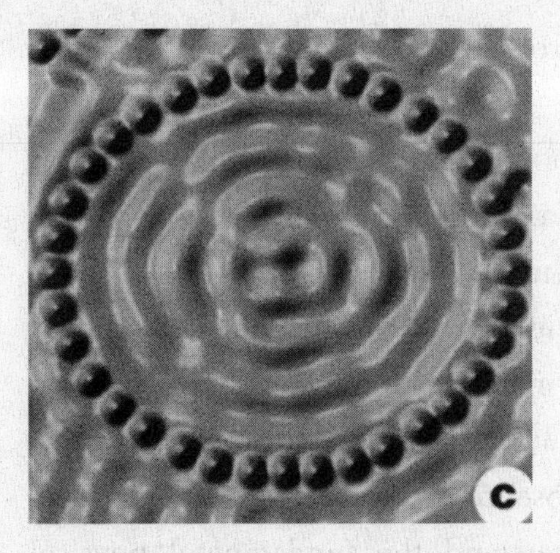

Imaging of atoms on a surface using scanning tunneling microscopy. (Courtesy of *Nature*)

cultural too. These instruments are expensive, costing in the many millions of dollars. Each institution must have at least one, but certainly not every research laboratory within the university or government or corporate laboratory has its own. These facilities, and others like them, are shared. They become meeting places for the researchers that use the instruments. And the user base is diverse: biologists wish to image the inner structure of mitochondria; chemists, the patterns of atoms within their nanoparticles; materials scientists, the conformation of nanocrystals conferring strength on structural steel. Common tools provide a place of convergence for researchers spanning interests.

I recently had two technique-driven encounters of different characters and outcomes. Both occurred when I was using an atomic force microscope—a cousin of the scanning tunneling microscope—to examine how quantum dots had self-organized into distinctive patterns on a glass slide. The first encounter came when a researcher from another laboratory saw my image and immediately realized that we had a shared interest in the organization of quantum dots on surfaces. This led to a discussion of methods aimed at improving our control over the nanometer-thick layers that the quantum dots formed.

The second encounter was more remarkable: another investigator saw the same image from afar and thought that I was studying the organization of latex spheres—micron-sized particles—on a plastic substrate. I was not: my particles were 100 times smaller than his, made of an utterly different material, dominated by a different chemistry, and organized onto an entirely different surface. The fact that the images were, in the absence of context, indistinguishable, motivates the search to understand the fundamental physical phenomena that unite the formation of patterns within our disparate materials systems. Why do clouds look like cotton balls and sometimes hyenas? Certain molecules like trees? Atoms, quantum dots, and latex spheres like the surfaces of egg cartons? A highly visual approach to the structure of matter, enabled by new methods in microscopy at the most refined resolutions, reveals the common threads between erstwhile disconnected patterns. That scientists are drawn by aesthetics pulls their minds and their visions together from across vast intellectual landscapes. As Harry Kroto said on positing that the buckyball, the European standard soccer ball, and Buckminster Fuller's geodesic dome were one and the same shape, "It was so beautiful it just had to be right."

Learning to Write

From Jean-Marie Lehn's principles of molecular sociology one should—with infinite information and understanding—deduce everything. Atoms, molecules, molecular interactions, recognition and organization, through layer upon layer of complexity ultimately leading to all things living, and everything else too. To set as our goal to emulate life's creation of functional complexity from atomic simplicity sets an even more ambitious project than Nature's own. She did not plan, but instead rolled the dice, and selective pressures that propagated certain variations did the picking and choosing. The nanotechnologist wishes instead to specify an outcome—a solar-to-electrical energy-converting wall ivy, a bulletproof garland—and invert this objective, tunneling down to the molecular seed that will engender it. Today it seems a daunting challenge to synthesize, from the atom up, molecularly programmed systems of the complexity that we ourselves embody.

More compatible with our current notion of order and planning is to think of designing from the top down: to specify function and then plan its execution. Our success in building computers proves that we are adept at the design and planning of a complex system orchestrated from above. Today, new computer chips, now with billions of interconnected transistors on each one of them, are issued every eighteen months. Electrical and computer engineers design and build new editions of these logically interlinked letters, words, sentences, paragraphs, and chapters with exquisite control. Today the transistors that make up these computer chips are on the verge of being made smaller than 100 nanometers (see Chapter 7, Compute).

The computer architecture problem is itself hard enough to solve. Like authoring an intricate narrative, it requires a systematic approach, a vast memory, and an eye and ear for detail: one must map and track the comings and

More compatible with our current notion of order and planning is to think of designing from the top down: to specify function and then plan its execution.

goings of *Anna Karenina*'s Varenka, Vasenka, and Varvara Vronsky without conflation. The ability to erect and link these functional building blocks so abundantly, so reproducibly, and with such complex interconnection has been critical to our mastery over computing and information technology.

In computing, the letters of the alphabet, the smallest unit of currency that the engineers deploy, are not atoms or molecules, but transistors. One electrical contact controls the flow of current between another pair of electrodes. Transistors in this way amplify signals in radios and form switches in which one terminal controls the others. Cascading these transistors leads to complex logic circuits inside our computers.

Vacuum tubes, predecessors of the transistor, performed a similar function. As amplifiers they enabled home electronics. These were essentially light bulbs with a third terminal that controlled current between a pair of electrodes. Lee de Forest, an American inventor and physicist, made the vacuum tube triode that served as an amplifier for radio signals and thereby made possible the AM radio. The vacuum tube triode advanced computers as well, and was deployed in building the early systems of the late 1940s and early 1950s. These triodes led to more powerful computational engines than ever had been seen before. But vacuum tubes tended to leak, and the metal that emitted electrons in the tubes burned out. They required so much power that increasingly complicated circuits took too much energy to run. Computers built in the late 1940s occupied 1,000 square feet of space, and machines with 1,000 vacuum tubes were unreliable and power-hungry.

In the 1920s scientists had begun working towards another approach to amplification and logic, built not on glass tubes but on solid crystals. Devices with two metal terminals—a sharp metal tip contacting a piece of semiconductor crystal—formed diodes, devices at the heart of crystal radio sets. In 1947, John Bardeen and Walter Brattain, working at Bell Telephone Laboratories, were investigating how electrons behaved at the interface between a metal and a semiconductor. They made two point contacts close together on a semiconductor, producing a transistor instead of simply a diode. Connecting a number of these together, they realized an amplifier. Unlike the vacuum tube circuits of the time, this solid-state amplifier didn't need to heat up. The invention of the transistor just before Christmas 1947 marked the start of modern semiconductor technology, and won William B. Shockley, John Bardeen, and Walter H. Brattain the Nobel Prize in Physics in 1956. Transistors were smaller, more reliable, and more energy efficient than vacuum tubes, and they allowed more complex systems to be engineered. Soon tens of thousands of transistors were being connected together, increasing the complexity of computers.

These machines were still being built rather cumbersomely, however, with individual components soldered together, making further growth in size a daunting challenge. The situation resembled pre-Gutenberg book production, each additional copy of a circuit written out longhand with monastic patience. What was needed instead was a means to build and connect transistors all at once, and print them over and over like letters on a printing press.

In 1958 and 1959, Jack Kilby at Texas Instruments in Dallas and Robert Noyce at Fairchild Semiconductor Corporation found a solution: a way to produce many transistors on a single circuit platform, all in one go. This was the birth of the integrated circuit. The approach was scalable: if 10,000 transistors could robustly be printed and interconnected, then so might 20,000, at little additional cost and effort. Especially if smaller and smaller transistors could be made, and if their performance improved when their size was decreased, then increasing the compactness and complexity of circuits would lead to a virtuous cycle of exponentially growing computational prowess and sophistication.

The resultant trend was named Moore's Law, after an observation in the 1960s by Gordon Moore of Intel Corporation that the number of transistors packed into a given area was doubling every eighteen months. Jack Kilby won the Nobel Prize for the invention of the integrated circuit, an innovation that led to decades' worth of subsequent progress, including personal computers, digitally switched telephone networks ... and spam.

The runaway success of silicon chips also branched into novel semiconductor physics and devices. Exotic semiconductors have given us microwave-frequency cell phones and lasers to light up fiber-optic communications. In these semiconductors, complex superstructures made from semiconductor crystal are built up, and electrons flow through an intricately tailored obstacle course. The promise of such extreme control over electrons led Herb Kroemer, co-winner of the 2000 Nobel Prize in Physics, to work out in 1957 the principles that would allow transistors to operate at 600 billion cycles per second, 100 times faster than ordinary transistors. Zhores Alferov, Kroemer's co-laureate in 2000, developed semiconductor lasers that poured, and then trapped, enough electrons into the light-producing heart of the device to make it amplify bright beams of light. These lasers—and their compactness, focused beams, purity of light, and low cost—have been essential to the fiber-optics revolution in communications (see Chapter 9, Convey).

While these intricately layered semiconductors look dramatically different from molecules such as buckyballs, nanotubes, and quantum dots, and they are made through a vastly different procedure, they showcase the same physical phenomena. They provide control and enable purpose-built engineered matter at the nanometer scale, remaining to this day our most refined means of tailoring matter to influence the flow of electron waves.

Choreographing the Dance of Molecules

We have seen that it is possible to design, build, and visualize matter on the nanometer lengthscale. Now we must choreograph a dance of molecules to harmonize with our corps de ballet's innate talents and tastes, and one that yields results of which we can be proud. How better to highlight nanotechnologists' approach—building from the molecule up—than to measure the march towards useful application? In *The Dance of Molecules* we identify central challenges in human health, environment, and information, and follow progress towards nanotechnologists' purpose. We explore the territory of today's and tomorrow's nanotechnologies in search of molecular choreography serving society.

Health

Each of us has seen a friend or loved one die too young. We each know people whose enjoyment of life's pleasures has been curtailed by sickness. Maybe we suffer this ourselves. In talking with family and friends, I've found that nanotechnology's role in medicine is at the top of people's minds. North America is investing $6 billion in science and engineering research in 2005 and over $30 billion in health research. Newspaper headlines tell the same story: cancer and osteoporosis are "cured" at least once a week on the front page of national newspapers. We have an insatiable appetite for reassurance—for many the matter is pressing—regarding illness, its cost to the length and quality of life.

We need to begin by improving diagnosis. In nearly every ailment for which we have a cure—partial or complete—early diagnosis vastly increases our chances of extending life and its enjoyment. Improved methods of diagnosis will combine sensitivity and specificity: they will detect disease when the subtlest signs exist, and will not raise false alarms that overtax our health-care system and distract it from more urgent problems. Once a disease is diagnosed, it can be treated. Improved cures will also be specific: if we are treating cancer, then we wish to attack the cancer cells, and not hair and intestinal cells. If our therapies can be specific, then we can work at high doses of drugs—doses lethal to cancer cells, but no longer lethal to the patient if designed to attack only where they are needed. A third goal in medicine—a bold extrapolation from conventional curative medicine—would be to replace failing organs with new ones before the situation reaches a crisis. Today, organ transplantation is a predictable, successful science. The problem: there are not enough organs available. What if, instead of waiting for donors to come along, we could grow organs outside the body? Hydroponic tissue farms.

Nanotechnologists, by manipulating matter from the molecule up, are sowing the seeds of medical revolutions to come.

1

Diagnose

Cancer occurs at the nanometer scale. It is a mistake in a molecule, an error in the DNA that programs our destiny. Slip-ups the size of nanometers have huge human consequences.

Inside we are abuzz with molecular mechanics. Whirring motors suck away sugar's stored energy. Transporter molecules drag heavy loads along nanometer-narrow girders spanning the insides of our cells. Dangerous bacteria seek to colonize our bodies and we produce molecules to resist invasion. As with David and Goliath, our 10-nanometer molecules use their smarts to outwit attackers ten times their size. Romance happens too. That two molecules were destined to be together, bound to one another, cannot be determined at a demure distance by checking out each other's incomes, aptitudes, and family histories. Instead, the cell is a vast orgy, molecules trying one another on for size with surprising promiscuity. Molecules fit into one another, or don't, depending on size, shape, and affinity.

Molecular keys fit uniquely into particular locks. When a key and a lock recognize one another, they unleash a whole sequence of life-critical activities: the feeding, reproduction, or death of a cell. Protein molecules begin as simple chains, but chains with folding instructions that, like origami, produce a striking variety of shapes when Nature carries out the instructions. The resulting complex structures mate only selectively. DNA contains the code describing the chain of molecules that will make up a protein, and through this it determines the shape and function of the resulting folded protein. The translation from DNA into protein is a word-for-word transformation: English to Pig Latin. Our DNA is a complete library that describes every protein we need. From it limitless copies can be made to produce the billions of identical protein keys of each type in daily use inside our cells.

Within this buzz of activity, something is bound to go wrong every once in a while. There can exist both innocent and consequential errors. The machines inside our cells that manufacture proteins from DNA instructions can make a mistake and it will matter little: a useless key floats around, fitting nowhere. A mistake in copying the contents of the DNA library while cells are replicating, however, will alter the set of proteins the cell will produce. If the cell cannot survive without the right protein, it will die. If the mutation changes how the protein, and through it the cell, works, but not catastrophically, then a fortuitously evolved form of life is created—or a new disease spawned.

To live means to make mistakes: being open to fortuitous variation means being vulnerable to danger. If we cannot avoid mistakes, then we would do well to learn to detect them early and act on this information promptly. The opportunity for early detection could transform how we deal with cancer. According to David Ahlquist, professor of medicine and director of the Colorectal Neoplasia

Clinic at the Mayo Clinic in Minnesota, a healthy colon lining becomes a polyp and eventually cancer over a span of seven to ten years. Instead of waiting for a polyp or the cancerous tumor to become apparent as a tumor colony containing one billion cells, what if we could see cancer when it is a molecule, or perhaps a few cells? That people must inevitably die from cancer may turn out to have been a fallacy derived from our current place in history, as erroneous as similar thinking during earlier times of plague or tuberculosis, and we may then look back on our beliefs about medicine as primitive.

That people must inevitably die from cancer may turn out to have been a fallacy derived from our current place in history.

Nanotechnology can help in this quest for early diagnosis—before cancer spreads, before Alzheimer's takes hold. Aided by chips that merge computer technologies with cells and genes and proteins, researchers are beginning to inspect cells one by one. With the aid of nanotechnology, medicine may protect its patients earlier: instead of looking only for massive tumors, we are gaining the ability to inspect each cell and thwart the malicious before they run amok.

Lighting Up Cancer Cells

In conventional medicine today, we use imaging techniques such as computed tomography (CT) and magnetic resonance imaging (MRI) to look for suspicious structures, tumors the size of a small grape. When it reaches this size, the tumor already contains one billion cells; these cells have undergone at least thirty generations of reproduction, giving them ample opportunity to spread throughout the patient.

In the summer of 2004, Professor Shuming Nie and his team at Georgia Institute of Technology used nanotechnology to look at individual flags on the surface of cells, and to light up only those cells that betray cancer. They reported that cancer might in the future be detected when it reaches only 10 to 100 cancer cells—a factor of ten away from diagnosing cancer in the first malignant cell, and thousands to millions of times more sensitive than detecting tumor masses once they have already developed to the size detectable by a CT or MRI. Nie built nanometer-sized beacons that shone brightly, announcing their whereabouts. He then put highly selective molecules on the beacons' surfaces—Velcro that would stick only to cancer cells. By injecting his beacons into mice and looking to see whether his beacons had stuck, Nie could tell where cancer was developing.

Nie poured many elements of molecular design into his beacons. These needed to be small enough to course freely through the bloodstream, a requirement that prompted him to build probes that were about 10 nanometers in diameter. Each 10-nanometer beacon was itself multilayered: gobstoppers, nano matryoshka dolls. The outermost doll would be the bait, making the probes attractive to cancer cells alone, enticing them to decorate themselves lavishly in the gaudy garb of the flashy beacon: gold to a rapper, white socks and Birkenstocks to an academic. At the heart of the probe was the light-producing beacon itself, a sphere of semiconductor 5 nanometers in diameter, such a small speck of matter as to be called a quantum dot. The color of light it produced, and the purity of this color, would be of profound

At the heart of the probe was the light-producing beacon itself a quantum dot.

importance. A characteristic hue would allow it to be seen against the undistinguished glow emanating from all flesh illuminated by a laser. Nie would engineer his quantum dots to produce as pure a tone as possible, and thus stand out against their background.

In Newton's world of classical physics, the atoms that go into making a piece of material determine fully the color of light it will produce. It's chemistry alone, and not the size of the particle, that matters. In the quantum world, however, the rules are different. When nanometer-sized objects are to be built, both the atomic constituents and the physical size and shape of a particle are important. The electrons that orbit atoms were declared by modern physics a century ago to be wafting waves of probability. Like guitar strings, we may tune electrons' lengths and frequencies by changing their length. By controlling the size of his quantum dots—his beacons—Nie engineered particles that produced light with a distinctive orange-red hue. In this way he created contrast, allowing the nanoparticles' bright orange splash to be readily discerned against the indistinct background of the mouse's own glow.

Nie then set about attaching designer-molecular Velcro to the outer surfaces of his beacons. The outermost layer, the one seen by the mouse's cells, would be made out of proteins, molecules engineered to stick only to particular cells: those showing signs of cancer. Previous researchers had already found the molecular key that fit particularly well into locks on the surface of human prostate cancer cells. Nie would decorate his beacons with this specially designed fluffy Velcro chosen to stick to the cancer cells. The beacons would announce only the cancer cells to be detected.

The quantum dots and the sticky proteins were so different from one another that Nie had to find a way to bind them together securely. He first wrapped his nanoparticles in a dense foliage of molecules pointing outward from the surface of the semiconductor particle—using nanometer hairs to make a furry quantum ball. One end of each molecule was designed to adhere strongly to the dot, the other to keep the dots well separated from one another. These spacers would keep the quantum dots from clumping together and losing the distinctiveness of their light-emitting properties, a behavior that arose from their precisely tailored size. The molecules with which they capped the surface of their dots created an insulating lifejacket, helping to keep the quantum dot afloat—dissolved in the bloodstream—and to preserve safe within the quantum dot the energy that would be injected to make it produce light. The brightness of the quantum dot probes—and thus the detectability of cancer in as few cells as possible—depended on the design and robustness of these capping molecules. To tell dim dots from tissue's native light-emitting properties would be like trying to distinguish a cream-of-peach color against an eggshell-hued backdrop.

Next in Nie's design was to stick on the double-sided tape that would allow him to decorate his beacons in cancer-cell-recognizing molecules. For these he used polymers—molecular chains composed of many links. Making this layer of molecules as sticky as possible required the researchers to design many anchor points for the specific-binding protein layer that was to follow. More docking stations meant more specificity: his beacons would stick to cancer cells and nothing but, and the culprits alone would glow brightly.

With the final protein layer added, Nie set about proving that his strategy worked. He first showed that, in a Petri dish, his quantum dot probes stuck only to human prostate cancer cells. By taking a known key and attaching it to the surface of his beacons, he had not compromised the specificity and strength of the previously designed tumor cancer-seeking molecules. Curving the fluffy half of designer Velcro around his dots left the molecules sticky and picky.

Nie then introduced the probes into the mice, using a syringe to inject his mixture into the tail vein. He illuminated the mice to power up the quantum dots, giving his nanoparticles the energy needed to produce their characteristic glow. He looked at his subjects using a camera and a filter that passed only the colors specific to his quantum dots.

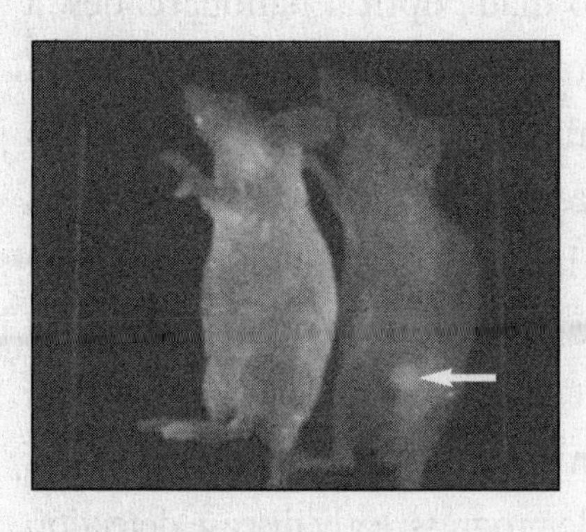

Two mice, the left healthy, the right suffering from human prostate cancer. Marked by the arrow is the location of bright orange light coming from quantum dots that have adhered to human prostate cancer cells. (Courtesy of *Nature*)

It worked: they detected the cancer cells. A bright orange light shone out from the sick mice, and its glow emanated from the point where they had injected human prostate cancer cells. If it can be translated into human patients, Nie's experiment could mean early diagnosis of cancer without surgery or tumor imaging. If surgery is necessary, Nie's method could mean "live," continuous, cancer-specific imaging throughout the operation, allowing the removal of the now brightly glowing tumor.

Nie set the stage for further improvements on his technique. He had chosen a particularly self-revealing form of cancer: human prostate cancer gives itself away by waving a single, distinctive pirate flag on the surface of affected cells. Not all cancers do. Many, though, do display a distinctive combination of signs, and Nie's experiment will allow researchers to look for such combinations. He designed his beacons to allow a few different markers to be attached, and he made his beacons color-tunable so that he could follow the whereabouts of multiple probes. Multi-marker probes could one day open cancer's more abstruse combination locks.

When many different markers have to be followed, the number of beacons that we need to distinguish will grow beyond the set of colors in our palette. Nie has also made optical nanobarcodes, a collection of complex, multicolored, multi-quantum-dot optical labels. In this way he expanded the library of codes from 10 to 100,000. With these, 100,000 different types of cancer-cell markers on the surface of cells could be hunted down at once. Imagine screening for prostate, liver, bone, breast, and lung cancers in a single test—and imagine doing it at an annual checkup so that cancer never had seven years to rampage through our bodies.

Though Nie has made remarkable progress, early disease detection using quantum dots still faces important challenges. His beacons, made using the semiconductor cadmium selenide, can be toxic to cells, especially when they are illuminated using intense light. On the positive side, Nie's polymer protection layer keeps the semiconductor surface isolated from the bloodstream. But it is not yet known exactly where the beacons end up inside the patient. Ideally, well-protected dots will be cleared from the body through the kidneys. The issue of possible toxicity, however, needs to be addressed head-on before quantum dots can be contemplated for human use. This will be a major hurdle to clear, first to satisfy researchers that quantum dots are safe, and then to gain doctors' and patients' confidence.

The broader importance of the work may not, however, depend on proving quantum dots to be nontoxic. Zooming out, Nie's research revealed that a shining beacon can illuminate cancer cells inside the body, and the hunt is on for even better beacons and improved Velcro. But even once a safe, efficient beacon is found, seeing cancer in mice does not prove that we can see it in humans. Our organs lie hidden beneath deep layers of tissue that visible light will barely penetrate. Light from a few quantum dots circulating in blood will already be subtle, and a layer of absorbing tissue between them and the camera will make resolving tumors deep within impossible.

But even once a safe, efficient beacon is found, seeing cancer in mice does not prove that we can see it in humans.

There is a solution, however. Light in the infrared—not visible to us, but just as real and powerful as visible light—travels more readily through tissue. Cameras and filters are available in the infrared as they are in the visible. An MIT–Harvard Medical School team has shown that beacons that glow in the infrared can be used to see the lymph nodes of mice and pigs during surgery. The surgeons, guided by their infrared camera, fully removed the quantum-dot-labeled lymph nodes and satisfied themselves of the success of their operation before it was over. These infrared quantum dots had, however, lost their light-producing capacity too rapidly, dimming as optical beacons. My group at the University of Toronto recently announced that it is possible to build beacons in the infrared that keep their brightness in blood plasma over days and weeks. We built particles directly on a biomolecule, DNA taken from a calf, to protect the semiconductor particles within a molecular wrapper. The particles glowed brightly and stayed bright in blood at body temperature. By using the structure of DNA molecules to program the growth of nanoparticles of just the right size, we managed to produce light that would flow deeply into live tissue.

All of these breakthroughs need now to be brought together. Already each team has broken through traditional boundaries that artificially slice research into chemistry, physics, biology, and engineering. Now teams spanning basic science, engineering, and medicine need to translate basic discoveries, with the tools of engineering, into the operating theater.

A visit to the doctor might one day involve swallowing a pill containing color-coded Christmas tree lights that stick only to renegade cells within nascently cancerous prostates, esophagi, and brains. Surgeons could ensure that they had fully removed emergent tumors without needing to remove excess margins of surrounding tissue. With such early diagnosis, we can envisage a world in which we do not give cancer the time to run rampant.

Sorting through Cells: Molecular Spring Cleaning

Shuming Nie's molecular beacons put nanometer-sized probes inside the patient. Reading these probes and extracting the results happens outside of the patient through the combination of lasers, cameras, and computers. To add to the set of diseases that can be diagnosed—new forms of cancer, or genetic diseases such as cystic fibrosis—new probes with new selectively sticky surfaces need to be developed.

Another approach is being pursued: put the lab inside the patient. Biochips, also known as the lab-on-a-chip, might one day enter into their subjects: colonize the patient rather than stand on the outside and observe, arms crossed. Just like the chips inside our computers, biochips can perform millions of precise experiments in the blink of an eye. Unlike silicon chips, biochips' currency is not limited to a single form—the flow of electrons representing digital information. Instead they deal with cells and their insides; the machinery of proteins that make life happen; and the genetic code that is DNA, which contains the assembly instructions for the machinery of life. Biochips go beyond the Spartan, pure, inorganic world of perfect digital logic, augmenting it with the liquid innards of cells: wet, slippery, organic life slithering atop a rigorously logical chip—the marriage of Beauty and Brains.

Biochips allow us to merge the narrow but impressive performance of computation with the admirable elegance and variety of biology. The union provides us with new ways to understand what goes on within ourselves. Life is complex, and in ways not fully understood, in many realms of size: sub-nanometer atoms, many-nanometer proteins, long but narrow DNA molecules, micrometer-sized cells, colonies and organs and people from the sub-millimeter to the many meters. The relationships among these lengths form a hierarchy without which life as we know it would not function. If each layer did not build on the next, we would be less sophisticated than fungus, our sentience a truffle.

Quantum dots touch on one or two layers in the hierarchy—binding to proteins that reveal cancer, clumping into groups the size of tumors. Biochips could span a greater range, which in disease is important. The nanometer scale is undoubtedly important: What happened in our genes that led to cancer? How have proteins been affected? Questions on a bigger size scale are important as well. Could a 100-nanometer virus be engineered to combat selectively the guilty cells? How does a cell's environment—including the society of surrounding cells—create pressure for negative behavior? Could the cell gone wrong reform its behavior if placed among the right crowd? Biochips could get us closer to looking at these interactions—and the cures to disease that they could provide.

Systems biologists study the relationships between nanometer-sized details and our own day-to-day experiences. How do genes and proteins result in coughs, orgasms, the spilling of coffee, the curvature of hips? If we could understand how human reality evolves out of the chemistry of molecules, then we might control our fate: engineer improved Greta Garbos, subjugate cancer, master depression, overcome or at least slow ageing. Or make our skin convert sunlight to sugar and our eyebrows consume dangerous greenhouse gases.

Experimental systems biologists fantasize about performing experiments that would help us understand how function arises from structure. Consider what we might wish to do to understand, with a narrow category of cells, how a number of signaling molecules influence the cell's fate. How do different proteins program cells to relax, reproduce, or die? The answer will depend on the cell's stage of life: whether it is in early childhood, adolescence, parenthood, middle age, or retirement. And the proteins will also influence the cell's stage of life. This is an example of feedback—loops of causality that circle back on one another.

All of these variables necessitate an array of experiments. Huge numbers of cellular subjects must be recruited and interviewed about their most personal daily habits. Identical cells would be placed in a separate compartment for individual study. Each compartment would be furnished with a personal webcam to monitor the habits of each cell. Systems biologists demand to witness nanometric intrigue, resolving new plot twists every trillionth of a second. Their voyeurism knows no bounds: details of fluids excreted in replication titillate. Cells' reproductive scatology beguiles. The cell's fluid samples are not generous; each one contains a trillionth of a quart of liquid. It is reasonable to ask cells to donate one-thousandth of this quantity in the name of science. Nanotechnologists are successfully analyzing femtoliters of goo.

Biology is not chemistry alone—it is architecture and mechanical engineering as well. A community of cell biologists attributes tremendous importance to the role of tension and compression in the cell's skin and skeleton. These give the cell structure and connect it to its neighbors. Cell biologists poke and prod their cells one by one using the tips of atomic-force microscopes, studying how cells behave when push comes to shove. The lab-on-chip community needs to make millions of such controlled stimulus–response interrogation rooms to witness the cell's response to the good cop–bad cop routine.

The territory to be explored within each cell is vast indeed. Imagine a group of tiny scientists donning micrometer lab coats and wielding nanometer syringes, performing their batteries of tests systematically and reporting their findings to central headquarters. The research needs to be automated, and systematizing millions of repetitive tasks cries out for the power and performance of computers. Nanosystems biologists espouse putting nanolabs-on-a-chip to

Cell biologists poke and prod their cells one by one using the tips of atomic-force microscopes, studying how cells behave when push comes to shove.

work around the clock to help them unlock systematically the remaining secrets of life.

Linking Life, Chips, and People

Labs-on-chips and nanometer probes link the logical and the biological—the computational and the sensational, and the dry and the moist. They open avenues to applying our most powerful human-made engines of analysis to the most intricate, fascinating system ever engineered: the organism. With the human genome sequenced, we have read the string of letters that make up the book of life, but we know only a little of what they mean. We are at the earliest stages of decoding life.

When we do understand in detail how and why we work, we will unquestionably be able to intervene. There are at least two dimensions to the ethical future of biology understood well from the bottom-up: what do we do with the information, and what do we do with our newfound power to effect change?

On the matter of information: If quantum dots, labs-on-chip, and whatever else nanotechnology gives us can monitor and sort cells based on disease, and report back—say, over the wireless Web—then to whom shall they report? To your physician, presumably. Your insurance company too, though, might like to know about your early-morning mutation.

If we can begin to intervene in our own biology, where will it end? If we could rescue cancer patients, prolong the duration and quality of their lives, then we would not hesitate to do whatever it took. What else would we be willing to do, especially since less morally unambiguous possibilities will likely come to us before the curing of complex, dreaded diseases? How short does a prospective new child need to be before we call social disadvantage a disability? Will genetically preprogrammed nasal aquilinity be downloadable off of the wireless Web? Today we interfere abundantly with the course of life, heroically and without reserve. Nanotechnology will further increase our power to change the course of life. Our new powers will come with new responsibilities—and for this we shall rely on our individual and collective will and ethics, two areas in which scientists can provide information as to our growing capabilities, but citizens will have to add their resolve.

2

Heal

Seeing cancer when the first cell went bad would give us the best imaginable early-warning system. What would we do with the advance notice? Nanotechnology will allow us to act on this information, and to do so with precision and strategy.

Today's chemotherapy treatments for cancer illustrate the lack of focus and control we have today in delivering drugs where and when they are needed. Traditional therapies don't just attack cells in the region of the tumor—they attack cells everywhere. Unlike treatments such as radiation therapy, which focuses on specific tumor regions, chemo strikes the entire patient. Chemo is not so much like highly targeted smart-bombs as it is like carpet-bombs, with all of their collateral damage: hair falls out, fingernails are lost and intestinal symptoms are suffered. Second, conventional chemo does not seek out specific cells—ideally, only cancer cells—to apply its toxic effect. It does target rapidly dividing cells, and this certainly includes cancer cells, but also the cells that produce hair and enable the normal functioning of the intestinal system. Third, chemo today involves periodic injection of drugs and their subsequent decay over time until the next injection.

The effectiveness of a drug is determined by its concentration in the bloodstream over time, and conventional methods of delivering drugs do not allow us refined control over this critical aspect of therapy. Consider instead what could be: could we build a system to deliver drugs that sensed when and where inside the patient a drug was needed, and to deliver the drug molecules accordingly? Our cues as to where illness has struck, and exactly what form it takes, will be molecular. Nanotechnology, sensing and acting as it does on the molecular scale, has a role to play.

Controlling Drug Delivery in Time and Location

MIT professor Bob Langer is a legend, not only in research circles but in entrepreneurship as well, having taken biotechnology innovations from the lab to patients. He holds 400 patents that are licensed to eighty companies. Thirty products, either on the market or undergoing the FDA approval process, have come from technologies developed in his lab.

Many of his innovations surround the delivery of drugs when and where they are needed, and much of his work applies to cancer treatment. Delivering drugs locally has huge advantages in treating certain types and stages of cancer. Cancer drugs are toxic, and if the drugs can be delivered locally, attacking only the tumor rather than the entire patient, Langer's methods can reduce side effects dramatically. At the same time cancer therapies can be made more effective, since concentrations of lethal drugs that would kill the patient if delivered throughout the body can now be delivered directly to the tumor.

At least seventeen such local drug delivery systems are currently in use to treat a variety of diseases: advanced prostate cancer, brain tumors, and leukemia. Drug delivery systems such as these are used by hundreds of thousands of cancer patients each year. Transdermal patches for smoking cessation (Nicoderm, Habitrol) are other examples of simple delivery systems that control release over time. Drug delivery across the skin is also used to bring pain relief to cancer patients at a constant rate.

Langer, with Judah Folkman at Harvard University, in the 1970s addressed one of the early challenges in delivering drugs at a fixed, slow rate in time. One of the problems in delivering drugs periodically by pill or injection is that the molecules in the drugs decay rapidly inside the patient. For many therapies, constancy of concentration is most clinically effective, yet the level of the drug in the patient's bloodstream is not uniform over time.

Langer and others had been working on methods in which drug molecules were trapped inside wafers made of polymers, the long repetitive molecules that make up plastics, rubber tires, and the carbohydrates in food. Embedding molecules inside these solid materials, rather than injecting free molecules into the blood, slowed the drugs' release to the patient. The drug molecules leached out slowly and controllably in time. The problem was that, when applied to the larger molecules used in some pharmaceuticals, the drugs didn't escape at all—they simply wouldn't budge. Langer and Folkman devised new chemical methods to get drugs into sponge-like networks of polymers containing molecule-sized pores. Like groundhogs traveling through a connected maze of tunnels, the drug molecules would travel through a network of connected pores. The researchers controlled pore sizes and therefore the rate of release: they could dial up a release time lasting from days to years by controlling pore size down to the molecular scale. These systems are used today to treat advanced prostate cancer. Administered once, the implants last one to four months.

Langer's drug delivery vehicles can be used to control release not just in time, but also in space, bringing drugs to different parts of the body. Chemotherapy's side effects come from the fact that cancer drugs kill many cells, not just malignant ones. Chemo is administered at a lower dose than would best destroy the cancer cells, lest the cure kill the patient; to make cancer drugs more effective and safer, we would deliver a drug at higher concentration but only locally. Such a technique is now in use in treating brain cancer. The surgeon removes as much of the tumor as possible during the operation, then places small, slow-release polymer drug wafers at the surface of the brain where the tumor had been, lining the cavity. In one clinical trial, 31% of brain cancer patients were alive after two years, whereas only 6% receiving conventional brain tumor therapies survived. The FDA approved the approach in 1996.

Langer's drug delivery vehicles can be used to control release not just in time, but also in space, bringing drugs to different parts the body.

Designer polymer molecules have also been used to deliver drugs where they are needed, but now through the bloodstream. The method takes advantage of the special preferences of tumor tissues for molecules of a particular size. Tumors have leaky blood vessels, whereas healthy tissues are fed by structurally sound, leak-proof vessels. Small molecules used in traditional chemotherapy drugs can travel directly through the walls of blood vessels—healthy and leaky—with no preference for healthy versus cancerous tissues. Polymers, much bigger molecules to which it was possible to attach cancer-attacking drugs, seep through the walls of leaky, tumor-feeding blood vessels, and are swallowed up preferentially by cancer cells. Attaching cancer drugs to large molecules has another advantage: drugs remain in the bloodstream for longer, stretching the time over which the targeted therapy is active. This method of homing in on cancerous tissues is strikingly effective: researchers found that, in mice, seventy times more of a cancer drug attached to a polymer accumulated in skin cancer cells than in regular cells. Researchers found that, because the drugs accumulated so much less in healthy tissues, mice tolerated ten times higher dosages of polymer-attached drugs than of the free drug. In 2003, ten different such polymer drug systems were undergoing clinical trials in human patients.

The smart-bomb idea for drugs can be made even more literal. Liposomes are designer bubbles that can be filled up with a payload of drugs that can be carried through the bloodstream towards the target. About 100 nanometers in diameter, they trap drugs inside a secure membrane for safe transport. This is the bomb—how about the smart? Liposomes can be decorated with molecules that bind specifically to markers on the surfaces of cancer cells—proteins that cells display when they go malignant. The particles stick to cancer cells, at which point the liposome's membrane fuses with the much larger cancer cell. The liposome empties its contents within the targeted cell, injecting a massive dose of drug. Liposomes have been approved to treat Kaposi's sarcoma, which is associated with HIV.

Liposome smart-bombs need to overcome the body's own built-in missile defense system. Each of us has cells circulating that clear particles from our bloodstream, putting structures such as liposomes out of commission. While this is usually a useful function in protecting us against invaders, with liposomes it would prevent the smart bombs from making their delivery. Researchers have successfully added molecules to the surface of liposomes to prevent them from being attacked by the cells that would normally disable them.

Scientists are also figuring out how to put cancer cells out of commission by cutting off the cells' supply of nutrients. As a tumor forms, new blood vessels grow to supply it with blood. A hormone within us promotes the growth of these vessels. Drugs have thus been developed to inhibit new blood vessels, and modified drug delivery has again amplified the success of these drugs: when injected directly into the bloodstream, drugs are not effective, but when deployed using sustained-release polymers, they are. Released using a polymer drug-delivery vehicle and combined with conventional chemotherapy, such drugs have increased the rate of survival among colon cancer sufferers.

Drug delivery controlled by engineering molecular sponges has proven itself in the lab, in clinical trials, and in countless patients. The superiority of these methods over traditional injection proves the critical importance of controlling delivery in time and space. There are, however, certain types of problems that these chemical attachment methods do not, on their own, address. One of these is the delivery of treatments to the center of a solid mass of tumor, tissue poorly fed by the bloodstream and thus to which drugs will not readily travel on their own. The other class of drugs not susceptible to the methods described are those that become less effective when they are attached chemically to a large polymer molecule. The striking success of controlled release, combined with these challenges, has spurred researchers to pursue new methods of controlling how drugs are delivered.

Controlling Drug Release Using Implanted Microchips

Researchers have recently shown that they can use electrical signals from microchips to stimulate the release of drugs. With these, a veritable cocktail of drugs can be delivered, with the exact dosing and sequences in time all controlled with the finesse of a computer chip. These have been dubbed "pharmacies-on-a-chip."

A veritable cocktail of drugs can be delivered, with the exact dosing and sequences in time all controlled with the finesse of a computer chip.

The concept is to make many reservoirs on a chip, each one containing a known amount of drug. In early prototypes, pharmacies-on-a-chip were programmed before they were implanted, their schedules of delivery preset outside of the patient. In the future they could be regulated using control signals sent in from the outside—downloaded from the doctor's handheld computer, for example, or programmed by the pharmacist's cell phone. Each of us might hold a standard selection of pharmaceuticals implanted and provisioned to last for a year, with the usual array of antibiotics and antihistamines loaded up. With each of us carrying around a lethal load of drugs, the security of the communications networks that program these chips would become of even greater importance than it already is.

Control from the outside could be improved upon by control from within. Imagine a self-regulating insulin delivery system that sensed blood sugar and how it was changing in time, and released tiny shots of insulin inside diabetic patients instantaneously and on demand. Instead of treating separately the measurement of chemicals in the blood and the therapy in response, feedback loops of detection and action could be built, supplementing our own natural stimulus–response processes as needed.

Delivering a variety of drugs with programmable control over their release in time also lends itself to combination therapies. In cancer therapy, a first drug would cut off the supply of blood to tumors by preventing growth of the blood vessels that feed them. This would be followed by chemotherapy to kill remaining tumor cells. And the patient would then be maintained on long-term, low-level therapy designed to prevent further blood vessel growth in the tumors. Specific 1-2-3 punch sequences for maximum effect.

At MIT, Bob Langer, working with Professor Michael Cima, reported in 1999 a new success in controlled drug release. They built a silicon microchip that, if implanted into a patient, would release a variety of drugs on demand. First they constructed a series of individual reservoirs. They filled each one with liquids that in a future drug delivery chip would be the drugs to be released. Finally, they covered each one with a thin membrane that would dissolve only when an electrical signal was applied. These devices had no mechanical parts and were simple to build. Their chips are illustrated in the figure on page 57.

Each reservoir held a minuscule volume of drug: a few billionths of a liter. Sealing the drug in, on top of the reservoir, was a gold membrane. The researchers chose gold because it's known already to be safe inside human patients, and because it doesn't react chemically and resists corrosion. They showed that they could dissolve the gold membranes one by one simply by applying an electrical voltage, causing an amount of current to flow that could easily be delivered using a minuscule battery. Charged chlorine atoms present in the human body already were attracted to the membranes, and caused the gold roofs covering the reservoirs to react and then dissolve.

To check that their chip worked, the group filled up the reservoirs with a liquid that would glow. They used ink-jet printing to squirt a controlled amount of liquid into each reservoir. With this simple technique they had tremendous precision, controlling the amount of liquid in each reservoir with accuracies much better than one-billionth of a liter. They proved that the gold membranes were sturdy coverings that sealed the filled containers effectively. They managed to store their devices for over a year with no degradation.

The researchers put their device in a biological solution and applied a small voltage—less than that inside a watch battery—with the result shown in the figure below. After one minute elapsed, their gold membrane was gone and the drug released.

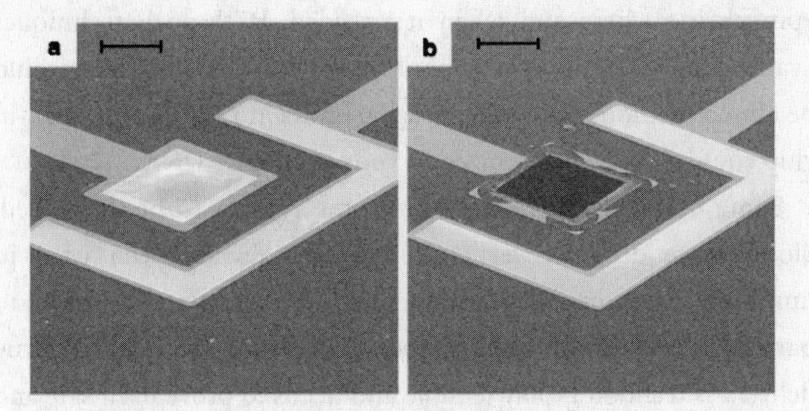

Removal of gold membrane resulting in release from the reservoir.
The scale bar is 50 micrometers long.
(Courtesy of *Nature*)

Langer's chip can easily be made self-sufficient. Powered by a microbattery, controlled using a circuit, with programs stored into memory, the device could be implanted to deliver its timed course of medication. Smart pills could sense their environment and act in accordance with their built-in program. If their programs tell them that something is amiss, they inform the patient and schedule a trip to the emergency room or to the family clinic, as appropriate.

The most versatile pharmacies-on-a-chip would not be loaded up with a limited set of pre-made drugs. Instead they could contain the molecular building blocks to enable an array of therapies. With labs-on-a-chip, built from microchannels, pumps, valves, mixers, and separators, each of us could contain a selection of full-time, implantable, personalized biotech start-ups. These could analyze chemicals in our bloodstreams, look at the state of DNA and proteins, and perform experiments necessary to discover the drug best suited to our condition. Headache? Let me concoct some Aspirin for you. Tired? I'll make you a nanocappuccino. Going into anaphylactic shock? Put away that monster needle; I'll release your epinephrine where and when it's needed. With such techniques available on-chip, the spectrum of drugs issued from within would be almost without limit. The drugs manufactured on your built-in chip could even be customized to your particular DNA.

Some of the tools are with us, some remain to be developed. Slow-release drug delivery systems have already been used in hundreds of thousands of patients, and liposomes targeted to particular cells are in clinical use as well. Microchip-based drug delivery is technologically feasible and needs to prove itself safe and effective in patients. To deliver drugs based on measured levels of chemicals in the bloodstream, safe, reliable, and long-lasting chemical sensors are needed. These exist for some important chemicals, such as oxygen, but for others, such as glucose, they often last only a few months in the demanding environment inside our bodies. The microchips must be powered too: microbatteries are effective but a lasting solution would be preferable. Techniques for charging up the systems are emerging, including radio-frequency and ultrasonic methods. Recent progress at the University of Toronto in making solar cells that are energized not with visible light, but with the infrared wavelengths that travel less impeded through the skin and tissue, offer another promising solution (see Chapter 4, Energize).

Detection of the molecules present inside us has also seen tremendous progress. Research teams such as that led by Professor Shana Kelley at Boston College, and that led by Professor Chad Mirkin at Northwestern University, have in recent years advanced the fields of DNA and protein detection at an astonishing rate. The goal is to see the signatures of disease, or clues as to cures, based on as few molecules as possible. The Kelley group has tailored the size and the shape of nanometer-diameter gold wires to control chemical interactions between molecules for maximum impact.

With these discoveries, medicine is coming closer and closer to inserting itself into the complex systems of chemical interactions at work inside our bodies. Technical prowess alone is not sufficient: we are blind if we cannot deepen our scientific understanding of cause and effect. By uniting the worlds of biology and computing, we are gaining further insight into Nature's methods from the molecule up; and nanotechnologists are sharpening their tools, ready for use as new knowledge is gained.

3

Grow

When your car's brakes or steering begin to fail, you go to the repair shop. Sometimes the part can be fixed; other times, it needs to be replaced entirely. Either way, there's no point in procrastinating, even if the spark plug isn't quite dead yet. Operating the car—a complex system with almost untold interactions among its many carefully engineered parts—could bring about new and even more costly problems. Must not the same logic apply to our own bodies? The argument for preventive maintenance of our own internal organs must be at least as strong, for what is at stake is so precious: our health. A failing liver impacts the entire patient. If repair cannot yield a fully functioning new organ, then isn't replacement a better option?

Organ replacement is now effective and widely practiced. Transplants from others produce excellent results, prolonging lives by decades and improving quality of life immeasurably. The problem is that healthy donor organs are in scarce supply. Their limited availability is, in many respects, good news: the use of seat belts has decreased the number of healthy young donors. But this comes at a price for a patient who desperately needs a new kidney. In North America more than 80,000 people are awaiting organ transplantation. A new person is added to the waiting list every twelve minutes. Only 23,000 patients received transplants in 2000. Some 15% of the potential candidates for liver or heart transplantation die while on the waiting list.

Is the only way to produce an organ for transplantation to grow it, over twenty years, inside a human subject? A kidney is, after all, nothing but a bunch of atoms correctly configured relative to one another. These atoms form molecules, and these are organized into cells which, collaborating in communities, coordinate to function as an organ. If nanotechnologists are not today equipped to grow an entire Greta Garbo from scratch, they are at least making progress towards bringing her pancreas back to life.

In North America more than 80,000 people are awaiting organ transplantation. A new person is added to the waiting list every twelve minutes.

Growing Replacement Organs in the Laboratory

Researchers in tissue engineering, or regenerative medicine, grow spare body parts in the laboratory. Their Holy Grail is to generate more abundant, safe, and compatible custom replacement body parts than what we have today through organ donation. The strategy is as follows. First, they decide on the size and shape of the organ. This can be customized to the patient's specific needs: for a replacement leg, the detailed structure of bones, cartilage, tendons, ligaments, blood vessels, muscles, nerves, and skin would all be measured and thus specified. The tissue engineer then produces a scaffold; like the inner skeleton of a building, this will determine the dimensions and general structure of the body part to be grown. Finally, the scaffold is populated with cells, after which it can be considered tissue, ready for implantation into the patient.

The cells used in tissue engineering are themselves an exciting area of research. They are taken from the patient, a feature that avoids the incompatibilities of genetics or blood type that plague traditional transplantation. But if we are missing a kidney, then are we likely to have kidney cells sitting around ready to populate a scaffold? No. Instead, we will use stem cells—cells that start off generalists but can be cultivated both to grow in number and to specialize into particular cell types.

Creating the Right Building Materials

The scaffold must provide the architectural foundations on which cells will grow to form an organ. Drywall and kitchen appliances do not, on their own, make a home: an outline of architectural form is required. The scaffold defines the structure of the organ that will take on life when populated with cells. It must nourish the cells as they travel to the innermost parts of the organ. In fact it must meet all of the particular cells' specific molecular needs, like handing out Gatorade at appropriate intervals along a marathon route.

Long, repetitive molecular chains—polymers—are used to build scaffolds. Their molecules have to be defined to meet a number of criteria simultaneously. The scaffolds need to allow cells to travel through them, something that requires pores that give rise to tunnels of the right size and abundance. The walls of these tunnels have a great deal of surface area, providing the opportunity to place molecules along the cells' path to cheer them on and nourish them as they go. Finally, when these organs are implanted, the original artificial scaffolds are intended to disappear and eventually be replaced by our own natural structural materials. Artificial scaffolds should biodegrade, but not so quickly that they vanish before their purpose is accomplished.

The U.S. Federal Drug Administration has approved a number of classes of polymers for use in appropriate clinical applications in humans. One polymer degrades quickly, in two to four weeks, which can be too little time to grow an organ. Another lasts for years. Scientists have successfully combined these two polymers in various proportions to make designer scaffolds that last just the right length of time. Natural polymers already found inside us have also been used—or rather reused—as scaffolds. Collagen is one familiar example. Making up one-quarter of all of the protein in our bodies, collagen forms molecular cables that strengthen the tendons and vast, resilient sheets that support the skin and internal organs.

At the other extreme of hardness are hydrogels. These are human-made versions of natural gelatin, early generations of which are already widely used in the form of soft contact lenses. Like gelatin, hydrogels can start off in liquid form and then be transformed into a solid. Taking the place of surgery to implant a new organ, hydrogels could be injected as a liquid, filling up irregularly shaped defects in existing tissues before turning into solid form. The process of turning to solid can even be controlled by the physician, with the aid of light, fusing together the long chain molecules, giving a form to what was once a liquid.

Architecting and Constructing the Scaffold

Since our organs vary widely, differing in size, shape, structure, and in what functions they perform, the scaffolds that tissue engineers create to promote growth of replacement organs also vary tremendously. These scaffolds do, however, share a common purpose: creating appealing spaces for cells to inhabit. Nice big lofts with high ceilings and attractive furnishings, yet comfy and cozy at the same time. Roomy without causing agoraphobia. They create a welcoming environment tailored to the cells of interest: chintz and a cat for pancreatic cells, glass and brushed steel for liver cells.

Researchers have created scaffolds using textile technologies, weaving biodegradable polymer fibers to create fabrics. Such simple structures have been used to engineer cartilage, tendon, ureter, intestine, blood vessel, and heart valve. These textile scaffolds, however, suffer the disadvantages of being mechanically weak, fast to degrade, and limited in the control they have over cells' living environment. They are but temporary trailers for cells to inhabit. Will cells cultivated in trailer parks grow up to be Oprah? Or will they turn into Britney Spears? With these as possible outcomes, we cannot afford to gamble.

Researchers have created scaffolds using textile technologies to engineer cartilage, tendon, ureter, intestine, blood vessel, and heart valve.

Researchers can build homes that are both cozy and hardy. In one method known as phase separation, molecules can be induced to self-segregate in the same way that oil and water do not mix but instead separate. Two polymers can be mixed such that they form distinct regions of one type of polymer or the other, not both. Other combinations of molecules can be made to separate into one region abundant in polymer, the other sparse; the solid material that results is porous. Conditions can be found that produce pores of various sizes that suit particular cells well. Appealingly, Nature does all the work in these methods, with the sizes of cellular apartments determined by the degree of chauvinism and bigotry of the molecules forming their segregated communities. Some tissues, such as nerve, muscle, tendon, and ligament, look more like tubes than like spheres, and phase separation has also been used to make scaffolds that match cells' geometries. At the University of Michigan, Peter Ma has shown that he can grow a scaffold with tubes pointed in a particular orientation, and that this helps organize certain classes of cells to align in the same direction—as cells in muscle do to achieve a coherent contraction.

The price of delegating much of the work to Nature is that we surrender complete architectural control. Thus, to satisfy the Frank Gehrys of tissue architecture, researchers set out to show that they can, if need be, build three-dimensional structures pore by pore. At MIT, Russell Giordano and colleagues adapted the methods of industrial manufacturing to the needs of tissue engineers. They employed computer-aided design and manufacture, first using software to design three-dimensional structures. They then built the structure up layer by layer. They first put down a layer of powder, and then selectively ink-jet printed a chemical that caused the powder to stick together. The regions of powder that had not been sprayed by the ink-jet nozzle were then washed away. They built up a three-dimensional network the way an apartment building is erected, floor by floor—except that in this case, a floor's worth of unset cement was poured and then fused only where concrete was ultimately desired.

Giordano's method produced designer apartments, but overly spacious ones—hundreds of micrometers, much larger than individual cells. Cells prefer cozier premises. They are also sensitive to chemical cues provided by nanometer-size molecules. One recent proof of this point came from George Whitesides and Donald Ingber at Harvard University. The researchers tiled the floors of cellular compartments with nanometer and micrometer patterns. The size of these patterns, much smaller than the cells, were found to determine whether cells lived or died. Design appears to have a huge impact on the life and death of cells: for some, Bahaus is Eros; Pei, Thanatos.

Bringing the Scaffold to Life

Sam Stupp at Northwestern University in Illinois recently built scaffolds that bring us considerably closer to the ultimate in nanometer control over the habits of cells. Stupp worked with neurons, the cells that carry messages across our central nervous system. Today, when an individual's spinal cord is severed in an accident or injured due to disease, paralysis can result. Stupp's research gives hope that we could one day employ advances in the field of regenerative medicine to restore function to people with damaged nerves.

Stupp's idea was to engineer chemical cues into his scaffolds. He would use proteins to program stem cells to specialize into useful nerve tissue, rather than form useless and even harmful scar tissue. He would build such a scaffold and populate it with stem cells, then investigate whether the strategy had been effective. Stupp began with the stem cells inside our bodies that help to replace central nervous system cells lost in disease or injury.

Stupp designed one intricate molecule that would, with the help of Nature's propensity towards self-organization, do everything he

Stupp designed one intricate molecule that would form itself into a scaffold and foster the growth of the desired nerve cells.

needed done: form itself into a scaffold and foster the growth of the desired nerve cells. Stupp built molecules that would organize themselves into long, strong rods similar to construction-site beams and struts. He did this by making his molecules different at their two ends. One end, the one where the nerve-tissue-promoting protein lived, was designed to love water. The other was designed to hate it, and would rather huddle with masses of identical molecules, their water-hating ends pointing inward, shielding one another from water. This strategy induced the molecules to organize themselves into long cylinders whose cell-promoting proteins formed the outer surface of the rods, aimed at the tissues to be grown.

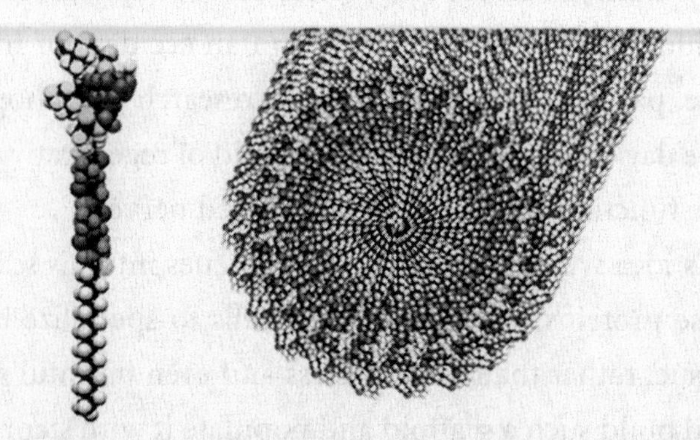

Sam Stupp's designer molecules and their self-organization. On the left is shown one molecule. Its tail (bottom portion) hates water, contributing to the subsequent self-organized formation of rods. The molecule's head (top portion) included a protein that promoted the formation of healthy nerve cells. On the right are shown the cylinders that self-assembled out from Stupp's designer molecules. The water-hating tails face inward and cluster together while those that love water face outward, projecting their cell-promoting proteins towards the tissues to be grown. (Courtesy of *Science*)

Stupp combined these self-organizing rods with a hydrogel strategy. His material would initially be a liquid, available for future patients to take by injection with a syringe. Before injection, the cylinders would be repelled by one another's negative charge. Only once the material was inside the body, with its compensating positive charges, would the cylinders be able to overcome their repulsion and assemble into a scaffold.

Stupp and colleagues added the stem cells, let the scaffolds assemble themselves, and gave the cells time to experience the influence of the scaffolds. Soon they realized that the cells were surviving well: nutrients, oxygen, and other molecules necessary to cell life were reaching the areas where they were needed. The researchers looked for signs that the scaffolds were helping the cells thrive, and found that the nerve cells grew much bigger when grown inside the active scaffold. They also examined the shape of cells, looking for clues that the cells were forming neurites—healthy, functioning nerve cells with branches reaching out to form connections with other nerve cells—instead of astrocytes, scar tissue cells. They found that the scaffold led to rapid formation of the healthy cells.

Stupp also investigated the role of the purportedly neurite-promoting proteins that covered the surface of his scaffolds. He compared two scaffolds, one covered in the protein, the other not. On the inactive scaffold, the stem cells didn't specialize into either good or bad nerve cells—they simply refused to mature. Still using the inactive scaffold, he added the protein to the mix, but not attached to the scaffold. The cells didn't integrate into the scaffold, nor did they sprout the neurites that indicate healthy nerve-cell formation. Stupp proved that both structure and chemistry are necessary to populate a scaffold with healthy nerve cells.

Turning Cells into Organs

What makes one cell grow up to make a kidney work, and another contribute to the functioning of a healthy liver? Researchers in the area of stem cells are answering this question and using their newfound understanding to build organs through tissue engineering.

Stem cells are freshmen that have yet to specialize; they have not declared a major in kidney function, bone formation, or sight. They have two important features: they create more stem cells without specializing, increasing the population of prospective cellular inhabitants of scaffolds; and they then differentiate into a variety of possible specialties—heart, kidney, or liver function, for example. In applications of stem cells in tissue engineering, researchers first harvest such cells; incubate them such as to increase their number; and then promote the specialization needed in the organ they are to populate.

Stem cells can differentiate into virtually any type of specialized cell. Progenitor cells, one level more differentiated than stem cells, are farther down the path towards specialization. They still have restricted capacity to specialize: they've chosen science over humanities, but have yet to pick one of chemistry, physics, or geology. Stem cells, since they can both multiply and specialize, are ideally versatile candidates to colonize scaffolds and thereby grow new tissues. They are useful beyond tissue engineering, too, being studied for use in heart disease, cancer, diabetes, and Alzheimer's and Parkinson's diseases.

Stem cells have been in the spotlight in recent years because of controversy over one of their sources. The most versatile cells of all are those from which all of our own adult cells are made: the fertilized egg, as well as the descendants of the first two divisions of this cell, can form any type of cell. After about four days, these cells begin to specialize, forming a shell surrounding the cluster from which the embryo will develop. These are pluripotent cells that can differentiate into any cell you or I could need, but they cannot produce an entire embryo on their own because they cannot create placenta and supporting tissues. Such cells can be obtained from embryos created for in vitro fertilization, donated with informed consent by the parents, and which would otherwise have been destroyed. Embryonic stem cells can also be derived from the tissue present in a fetus that would, when mature, develop into the reproductive organs.

Research in embryonic stem cells in mice has shown the amazing flexibility of these cells. In 1998, two published papers described the growth of human embryonic stem cells. These cells have since been shown to be capable of forming brain cells, skin, and functioning heart cells.

Therapeutic cloning provides another route to obtaining undifferentiated stem cells. The nucleus, the heart of the cell, can be taken from the patient's cells and transferred into a stem cell whose own nucleus has been removed. This technique employs the patient's own genetic material and therefore avoids the problem of genetic incompatibility between donor and recipient. However, therapeutic cloning followed by incubation has yet to result in growth of a population of more than six cells.

Many researchers believe that if embryonic stem-cell methods could be developed to work reliably, these techniques would provide flexible stem cells that could differentiate towards any specialization desired. However, in part because of concerns over how these embryonic stem cells are obtained, attention has recently focused instead on adult stem cells.

Scientists once believed that the sources of fully specialized worker cells were narrowly differentiated adult stem cells. These stem cells were committed to generating a limited range of progeny. The new view is much more encouraging: our adult tissues contain multipotential stem cells that can produce a number of different types of stem cells appropriate to their location. Stem cells located in our intestines can produce a range of different cell types needed in digestion and nutrient absorption. Adult stem cells, put into a new context, may even be able to transdifferentiate, turning into differently specialized cells not usually found in their traditional tissue environments. Some classes of the dozens of types of adult stem cells might even be as malleable as embryonic stem cells, and thus as useful in the field of regenerative medicine.

In tissue engineering, control over when and how stem cells differentiate is essential. Scientists recently have made considerable progress on this front. They know how to inhibit differentiation, a degree of control necessary to allow the cells to multiply in number and to ensure that cells do not differentiate towards a cell type other than the one that is needed. And they have found ways of culturing cells to make them differentiate into heart muscle, blood, skin, brain, nerve, skeletal muscle, bone-forming, fat, liver, and pancreas cells. Interestingly, cells respond to peer pressure: culturing them in the presence of mature cells drives them to differentiate towards the same type.

Tissue engineering is a tour de force of control of molecules and materials. It relies on cueing cells to respond to our wishes, and employs combinations of drugs and well-designed structures to do so. This reinforces a point broadly relevant to all nanotechnology: that cell biology is not all biochemistry, nor is it based purely on the mechanics of cells, nor is it cellular sociology alone. It is all three combined.

Implanting Engineered Organs into Live Subjects

Tissue engineering being such a complex and important task, it is critical to test its effectiveness in live subjects. These investigations began in the mid-1990s with studies in animals.

Bob Langer at MIT recently showed how tissue engineering could be used to tackle the biggest problem in modern North American health. Atherosclerosis—degeneration of the arteries due to the buildup of fatty deposits—leads to heart attacks, strokes, and blockage of the arteries to the legs, and it is the leading cause of premature death in North America. Surgeons bypass blockages in blood vessels using grafts, veins, or arteries taken from elsewhere in the patient's body. But many patients lack grafts available for transplant, for example, if they have required the surgery before. Langer saw the need to employ tissue engineering to grow materials outside the body to serve as vascular graft.

Bob Langer at MIT recently showed how tissue engineering could be used to tackle the biggest problem in modern North American health.

First Langer's group had to grow the arteries in the lab. They put smooth muscle cells into bioreactors holding polymer scaffolds. Langer hypothesized that the grafts would develop best inside an environment that closely resembled their eventual destination, and so he used a periodic pulsing, mimicking the flow of blood, to stimulate the vessels during growth. One set of grafts was grown under this pulsatile stress over eight weeks; a sample for comparison was grown identically except that no pulsatile stress was introduced. The team then implanted the grafts into pigs. The animals were monitored, and the only additional treatment was daily Aspirin. They found that the pulsatile-stress-grown grafts remained open and functioning for four weeks, whereas those grown without pulsatile stress showed blockages beyond three weeks. The team also found that the pulsatile-stress-grown grafts retained the sutures used to sew them in place better than did the others. Their technique had produced many cells that were no longer increasing in number, which was important in keeping the engineered arteries unblocked. Langer had proved that it is possible to cultivate arteries, aided by cells taken from the subject and grown on a scaffold to form useful tissue, and that this resulted in engineered tissue that served its purpose well when implanted into live subjects.

Nearby, at Harvard Medical School, Joseph Vacanti and colleagues recently tackled the problem of growing heart muscle in the lab. Fully differentiated cells are responsible for making the healthy heart contract. If these cells die during a heart attack, they cannot regenerate. Regenerative medicine offers a way to replace damaged heart tissue and restore a healthy heart. Vacanti and colleagues grew heart tissue by culturing, on a polymer mesh, heart cells taken from rats. These cells began beating after three days. The same team also recently reported progress towards creating a tissue-engineered stomach. They want to create tissue that would replace normal stomach tissue and successfully metabolize nutrients. The researchers took stomach cells and seeded them onto scaffolds; the engineered stomach tissues were then implanted into the mice. After four weeks, vessels began to feed blood to the stomach and a stomach lining started to develop. The researchers also saw evidence that cells were starting to form the muscle characteristic of the stomach.

Subsequently, in 2004, the same team showed they could produce bone for use in reconstructive surgery. Growing bone anew through tissue engineering would mean that doctors no longer had to remove bone from somewhere in the donor's body. Tissue-engineered bone would still be completely compatible with the patient because it is grown from the patient's own cells. Vacanti built tissue-engineered bone and used it, along with control samples, to perform reconstructive surgery. Six weeks following implantation, the engineered tissue resembled bone in hardness, and under X-ray examination, the connection between the native bone and the grafts was so strong and complete that the researchers could not make out the place where they had been fused. Blood vessels had even begun to develop in the implanted tissue.

Vacanti has recently grown new large intestine, crucial for patients who suffer colon cancer. In these patients, the colon must be removed and an artificial pouch introduced. The pouch is unable to take over the large intestine's role in processing nutrients, and the pouches can become infected, requiring further surgery. To grow a new colon, Vacanti took stem cells from the colons of rats and cultivated them on a scaffold. He found that a tissue-engineered colon could perform the physiological functions usually supplied by a healthy native colon.

The Future of Tissue Engineering

Engineered organs are a medical dream that is gradually coming true. New reports continue to emerge of tissue engineered, implanted, and integrated, its compatibility and function demonstrated in living animals. The potential of the field is vast: it could allow us to replace ailing body parts not just as they approach the point of fatal failure, but as preventive maintenance. Some approaches would necessitate surgery analogous to today's transplant surgery. Others, such as Sam Stupp's hydrogel-based tissue, could involve little more than an injection.

When a car accident or life-threatening illness strikes we may not have the luxury of six to eight weeks to wait for a new organ. Someday, as a result of tissue engineering, will each of us have a supply of spare parts growing in dedicated bioreactors at the tissue farm, ready and waiting for when we need them—a body double? And ethically, when will we feel we've gone too far? Without a doubt we would grow and implant an engineered kidney into a patient who would otherwise die. But what about upgrades? Bigger, better hearts for athletes. Leg extensions for aspiring models. Malleable new pieces of brain for those needing to master an entirely new trick late in life: learning to play chess, or mastering negotiation or the viola.

Today we often come down against performance-enhancing drugs, but we are not consistent. Overdoses of performance-enhancing caffeine are allowed—in fact pretty much demanded—in graduate school. Of course caffeine also enhances athletic performance, but the doping regulations of the International Olympic Committee recognize that it is a part of the normal diet for most people in the world. Thus athletes can take caffeine up to a certain level, and the permitted level is sufficient to provide a performance advantage. We may not like the idea of interfering chemically with our physiology if our ability is within the "normal" range. Yet already in the world today, the rich and the poor have vastly different access to the most effective performance-enhancers we know: the food we eat, including the right balance of proteins, vitamins, carbohydrates, and fats.

Amid this already complicated set of allowances and restrictions, what will we do with newly engineered bone, muscle, and cartilage? If we can develop procedures that minimize risk and keep cost acceptable, we may not be able to resist the temptation to experiment, to push the envelope. It is the natural extrapolation of our unquenchable thirst for self-improvement.

Amid this already complicated set of allowances and restrictions, what will we do with newly engineered bone, muscle, and cartilage?

With tissue engineering progressing rapidly, now is the opportunity to face these questions of principle and policy. We still have some time before our nanometer dexterity catches up with the possibilities we might imagine. Sam Stupp's work is among the first examples of our speaking persuasively to cells simultaneously in the nanometric language of biomolecules and the micrometric vernacular of scaffold architecture. And much work remains to understand fully how to cue the desired responses from cells and tissues. Many of the first engineered tissues have provided organs made of essentially a single cell type. Research is required to develop complex organs from many types of cells and substructures. The University of Toronto's Michael Sefton has set his sights on growing a heart, an intricately structured muscle containing crucially important blood vessels. A considerable chasm exists between what we can accomplish now and the engineering that will create such a complex organ. Furthermore, tissue engineering needs to be understood, and demonstrated as safe, inside human subjects, notorious for responding differently from animal models.

Still, we have come a long way in a short period. Langer and Vacanti have implanted simple engineered tissue in animal models, and these have been accepted, incorporated, and gained function within their subjects. The field's approach—using our own cells, ingeniously avoiding organ rejection—accounts for much of the rapidity of early success. Greater understanding of and control over stem cells may give us the power to return parts of our bodies to the earliest stages of growth, development, and rejuvenation. The era of youth regained—the smart, sexy, strong years, once thought long lost—may be nigh.

Environment

If our worries about our personal fate often center on health, then our global concerns focus on our natural environment. In particular we worry about how actions we take today will determine our children's future.

The question of energy springs first to mind. Our supply of stored energy, generously saved up for us in the form of fossil fuels, is finite. Moreover, when we liberate this energy, fossil fuels generate toxic pollutants and greenhouse gases. Our reliance on such energy and its providers is seemingly beyond our control. The United States, for example, with less than one-twentieth of the world's population, accounts for one-quarter of world energy consumption. Today it imports over half of its oil. The cost, over the past thirty years, of the United States's dependence on foreign oil has been estimated at $7 trillion, and the country's actions on the international stage are influenced by the practical reality of this dependency. How could the remaining superpower's behavior on the global political scene change if America were energy self-sufficient? The world may find out within one or two generations: nanotechnology is advancing our ability to harvest and store the sun's vast power cheaply and efficiently. At current consumption levels, America would need to cover only one-tenth of Nevada with efficient solar cells to satisfy the entire nation's energy needs.

Pollution of the environment—waste disposal, industrial production, and even deliberate terrorist action—creates a further complex of technological as well as political challenges. Research is already having an impact on this area by enabling the early, sensitive detection of environmental threats, the global analogue of the individualized cancer detection explored in Chapter 1, Diagnose.

Nanotechnology is providing a means of seeing threats before they become lethal. It is also enabling us to treat the environmental afflictions thereby diagnosed, reducing and even remediating the effects of pollution: cleaning up groundwater and reducing and treating harmful emissions. But there may be a downside to nanotechnology and the environment as well: nanoparticles may pose a threat to the environment and human health, and it is essential that we look at whether regulatory frameworks available to us today can anticipate and address unanticipated toxicities in new materials as they are created.

Nanotechnologists are not only applying their skills in aid of the biological, but are also taking inspiration from Nature in how they engineer new materials and functions. Researchers are learning from Nature in building nanomaterials that are active and alive, and also learning from natural machines such as the systems of proteins that translate sugar into movement and power. The field of biomimetics explores the links between the biological, the chemical, and the mechanical—the nanometer-scale connections we experience each day through the power of muscle, the hardness of seashells, and the resilience of flesh.

4

Energize

Each day the sun bombards us with ten thousand times more energy than we consume. If we could cover one-tenth of 1% of the Earth's surface with solar cells, and each cell was 10% energy efficient, we could satisfy our energy needs completely using this clean source of energy alone. If we could store and transport the captured energy efficiently, we could break the cycles that lead to oil spills, nuclear waste, greenhouse carbon dioxide, and noxious urban air.

For decades we have built efficient solar cells that convert energy from the sun into electricity. Solar cells are pristine captors of the sun's energy. They are photovoltaics, producing an electrical voltage and a flow of current, providing electrical power when they absorb light from the sun. Why, then, is our energy strategy fossilized in coal and not basking in sunlight? There may well be a political dimension—entrenched interests of the incumbent energy suppliers—but there is a practical reason too. The world is not solar-powered today because solar panels cost more to build and install than we are willing to pay.

The sun's large but finite intensity, combined with our ravenous appetite for energy, means that we need solar cells of large area to capture a useful amount of power. The surface area of solar panels determines the rate at which they can ingest uniform beams of photons, just as the enormous mouth of the whale shark captures more plankton than that of the sardine. We need to open wide. But the requirement for solar panels to cover large areas does not have to make them expensive and cumbersome. While one-tenth of 1% of the Earth's area is a lot to cover with optically absorbing materials, we've done it once already: America has covered or paved more than this fraction of its landmass. We might instead roll out solar carpets across the deserts of Arizona, paint the sands of New Mexico in order to harvest light.

Printing Solar Cells

Rigid and flat wafers of pure, perfect semiconductor crystals are used to make traditional solar cells. After decades of research effort and engineering, these silicon devices can now convert about 20% of incident power from the sun into electrical energy. If they were cheap and easy to work with, this efficiency would more than suffice. But semiconductor solar panels end up as big glass panes. The crystals need to be grown at high temperature in a pure vacuum environment. They have to be manufactured with extreme caution and cleanliness.

Now nanotechnologists are breaking away from large, perfect, crystalline semiconductors. Instead they are building physically flexible solar fabrics, as shown in the figure. These photovoltaics are printed like newspapers, spinning seamlessly from roll to roll. The method is scalable: if we can produce a foot of solar paper, then we can make a mile of it. The cells are easy to transport and deploy. And they could even be wearable: after all, a black turtleneck today absorbs enough light from the sun to power a light bulb or recharge a battery—why not turn this energy to electricity instead of heat alone?

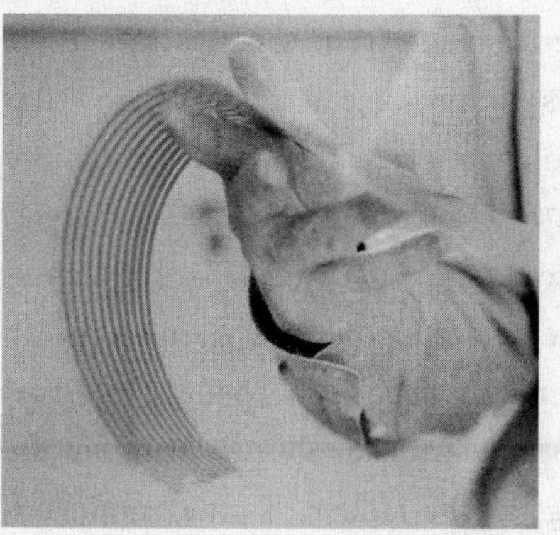

A plastic solar cell.

Electrically conducting plastics invented in the 1970s are the basis of one main approach. These are light, flexible, cheap, and easy to work with. Plastics are made from long molecular chains—polymers—which intertwined give them their combination of strength and flexibility. Once we associated polymers with car tires, Tupperware, and polyester: the essence of style and grace in mid-century North America. It was polymers' *non*-electrical properties that were prized: they were used to make the insulation coating electrical wires, taking advantage of polymers' stubborn refusal to conduct electricity.

Then in the 1970s Alan Heeger and colleagues at the University of California at Santa Barbara discovered the conduction of electricity

Plastic photovoltaics are printed like newspapers, spinning seamlessly from roll to roll.

in polymers. Electrons could be made to flow along the chains of certain long molecules. He coaxed Nature into building the smallest imaginable wires: less than a nanometer wide, these molecules could be made to act like metals or like semiconductors. The metallic molecules were similar to flowing mountain streams, and a material made from a network of these molecules merged the streams to produce a river of electrical current. The semiconducting polymers were more controlled, more restrained. Only certain classes of electrons, those having energies lying within specific ranges, could flow. They acted like vertical stacks of pipes, rather than like metallic polymers' deep rivers. To participate in the flow of current, an electron needed to be raised to the height of an available pipe. Electrons unable to surmount the gaps in energy separating rungs in the ladder of pipes could not participate in conduction. The semiconducting polymer was much more discriminating in its governance over electron flow than the metallic system. It gave researchers the control that would be needed to tailor light-absorbing properties to make plastic solar cells. The semiconductor bandgap—the flowless space separating conducting pipes—governs how a material interacts with light. Efficiently harvesting the power of the sun relies on extracting as much energy as possible from each particle of light, and a semiconductor with just the right bandgap has the potential to capture a maximum of that photon's precious energy.

A particle of light that contains enough energy can boost an electron up to a higher level. The photon vanishes, losing the energy the electron gained. Not only does the electron enter into a new, higher pipe from which it can conduct electricity, but also a space is created in the lower pipe, a hole that can also conduct electricity. The propensity for the system to conduct electrical current is thus altered by the absorption of light. Achieving conduction due to light is a step towards a solar cell, but the device still needs a battery to propel the electrons to freedom. It senses light but it does not harvest power. The optical detector is a voyeur, a castrato ogling the photon but under-equipped to seduce it.

A true solar cell devours light. It develops a voltage using energy provided by the sun and thrusts forth a flow of current. This is achieved by building a bias into the system: just as the side of a mountain slants, propelling water downhill, in plastic solar cells, two different metals are used to contact the light-harvesting layer, and electrons flock towards the one that is at a lower energy, producing the flow of current and the development of electrical power.

The same Alan Heeger who discovered electrical conduction in polymers contributed a major breakthrough in solar cells using nanotechnology. Until the 1990s, plastic solar cells were inefficient because electrons lost their energy before the energy could be harvested: photons had launched the electrons from their comfortable armchairs up to the ceiling, but electrons were quickly relaxing right back down into their La-Z-Boys before the increased energy could be harnessed. This undesired process of electron relaxation happens only if there's an empty armchair available: the excited electron will land in the cushion only if his wife is not already occupying the chair. Electrons belong to a family of particles named in honor of the Italian physicist Enrico Fermi, father of the atomic bomb. Fermions have an exclusive character: two cannot sit in the same chair at the same time. Because the process of excitation creates both an electron and a hole, empty armchairs abound when and where energized electrons are present. Pretty soon everyone will find a seat. Back in their chairs, the electrons have given up their energy in the form of heat transferred to the plush cushion. This warms the chair for the next electron that comes along and simply makes relaxation more enticing still.

Photovoltaic devices must instead extract excited electrons' energy before relaxation sets in. One solution is to put fans in the corner of the room to blow the excited electrons out as quickly as possible. Unfortunately, with an abundance of cushy chairs beneath them, the electrons are sorely tempted to flop down even in a strong breeze. As a result, in early plastic solar cells, only the energy contained in 1 in every 10,000 photons absorbed was successfully harnessed. Heeger took a different approach to solving the problem. Recognizing that comfy armchairs were too tempting, he introduced another, competing enticement to excited electrons: he furnished a living room not only with couches, but also with chandeliers. He induced his excited electrons to swing from chandelier to chandelier, aided by the breeze. Heeger's electrons remained excited and their energy could be harvested once they had swung their way out of the room. Heeger arranged the furnishings such that no chandelier was placed directly atop an armchair. By reducing thus the temptation for an electron to flop into the comfy chair, Heeger ensured that, even if the Tarzan-swing across the sitting room took a while, it did not have to compete with attractions of chair sitting. In this way he kept his electrons excited and tapped much more of their energy.

He kept his electrons excited and tapped much more of their energy.

Heeger's armchairs were his polymer; his chandeliers, buckyballs—soccer-ball-shaped nanometer-size molecules, each made from sixty carbon atoms. Electrons love to take refuge in buckyballs. Once excited in the polymer, an electron jumped onto a bucky-chandelier at the nearest opportunity—in less than a trillionth of a second. Heeger thereby improved photocells by a factor of one hundred. Now for every hundred electrons excited, thirty managed to swing-blow out of the room still energized.

Heeger's success also guided researchers towards their next breakthrough in harvesting photons even more efficiently. If chandeliers could be made so tantalizing, then was it strictly necessary to offer electrons a direct path straight out of the room? Or—with a luxury of time and space afforded by the creation of separate escape routes for electrons and holes—could solar cells be made in which electrons wandered through a tortuous maze towards their ultimate escape? Biology teaches the value of tortuosity: our intestines, for example, are textured, and as a result, although they are compact, their areas for chemical interaction are vast.

Tortuous paths for the absorption of nutrients in food inspire a similar means of extracting energy efficiently from excited electrons. In both cases the action is at the interface—digestive chemistry in our intestines, electronic energy extraction in solar cells—and an abundance of area promotes the process. Sponges, a car's catalytic converter, and water filters use the same principle.

Michael Grätzel borrowed this natural engineering strategy. In his research at the Swiss Federal Institute of Technology, he recognized that a flat solar cell would not be an efficient one when all the action was at an interface. He made a molecular photocell that relied on the electron-hole pair lying right at the junction between two materials. But if only one layer of his light-absorbing molecules lay across a flat interface, little light would be absorbed: it would be like trying to sop up Niagara Falls using a regular, non-extra-absorbent paper towel. Grätzel scaled the sponge/intestine idea down to the nanometer, making a device packed with dye molecules and thus absorbent to the sun's rays, but in which each dye molecule lay at an interface. Grätzel's device was nanotextured. His solar cell blew away past performance records for photocells based on organic molecules, leading to solar cells whose solar-to-electrical power conversion efficiency exceeds 10%. At this level of efficiency one-tenth of a percent of Earth's area would need to be covered by solar cells to meet our energy needs.

Grätzel's solar cells have much in common with the photosynthetic process by which plants turn sunlight into stored energy. Plants use chlorophyll, a dye that makes them appear green, while Grätzel used designer dyes. Why not go all the way and use plants' chlorophyll as the dye? Plants have had the advantage of millennia of evolution over which to optimize the efficiency with which they absorb photons from the sun. A spinach plant is efficient, churning out a great deal of energy relative to its size and weight. And—for the benefit of environmentalists—it is incontrovertibly green. Marc Baldo and colleagues at MIT recently purified the active light-harvesting ingredient in spinach. They mashed up ordinary spinach and isolated ten nanometer-sized proteins from the plant's cells. They found a way to unite normally wet proteins with the dry electronics needed to run the devices. Baldo and colleagues' spinach protein has yet to threaten Grätzel's dyes in its performance. Better yet, though, the two methods might be merged to make long-lived, efficient, and highly absorptive Grätzel cells sensitized using the power of Baldo's spinach.

In 2005 researchers conquered yet another important challenge in solar cells. Heeger's polymers and Grätzel's dyes and Baldo's spinach harvest a good proportion of the light they absorb, but they absorb the colors of light we can see with our eyes. These devices "see" only about half of the sun's power reaching the earth. The other half of the sun's power lies in the infrared—wavelengths invisible to us, but as real and powerful as visible light. In fact any warm object—even one not so hot as to glow visibly—emits energetic infrared rays in scaled-down mimicry of the sun's production of visible light. The outer edge of the sun has a temperature of about 5500° Celsius. In an early chapter in the history of quantum mechanics, the dramatically titled Ultraviolet Catastrophe, the relationship between the temperature of a hot body and the colors and powers it produced could only be apprehended correctly with the benefit of the insights of Max Planck and Albert Einstein. By positing the discrete nature of particles of light, in addition to its wave nature, they made sense of the sun's colors and power: a quantum leap for optical physics.

It was known then, even before the catastrophe was resolved, that a direct relationship exists between a hot body's temperature and the colors of light it produces. The sun glowing at 5500° Celsius produces a broad, yellow-white spectrum of energy whose maximum lies in the green. Plants absorb red light and reflect the green, a fact that accounts for the color of their leaves. Their chromatic strategy is a matter of speculation as to how differently colored bacteria won and lost prehistoric battles for light.

A furnace at 1000° Celsius—hot, but nothing like the sun—emits rays that lie mostly in the infrared. The U.S. Navy would like to convert nuclear submarines' heat created in reactor cores directly and silently into electrical power using infrared-absorbing photovoltaics, bypassing today's steam turbines. Inside a practical submarine reactor, or any other furnace for that matter, the furnace walls would melt at the sun's 5500° Celsius. But at the much more realistic 1000° Celsius, abundant infrared power is produced and ready for harvesting.

Therefore photovoltaics are needed that can capture not only visible light, but infrared light as well. In 2005, my group at the University of Toronto showed that we could turn infrared power into electricity in a paintable solar cell. We used quantum dots, particles of semiconductor designed to absorb specific colors of infrared light, and harnessed photovoltaically the previously unabsorbed wavelengths that lie beyond those that we can see. We tailored the nanoparticles' diameter such that the materials started absorbing anywhere across a vast swath of the infrared color spectrum. These solar cells could thus be tuned to the temperature of a furnace, a portable heat lamp, or the tail of infrared power coming from the sun.

Research in the area of large, flexible solar cells has conquered a number of challenges in making cells that are efficient and customizable. Researchers still need to show that their new designer materials can be made on the massive scale required if we are to make a dent in our consumption of fossil fuels. A number of companies have already taken up this manufacturing challenge. Such a mix of tremendous benefits and significant practical challenges facing flexible photovoltaics epitomizes the opportunity for nanotechnology in the coming few years. It needs to prove itself not only fascinating, beautiful, and performance-enhancing, but also irresistible. When manufacturability, economics, and human factors

together are taken into account, the fruits of nanotechnologists' ingenuity must remain compelling.

Releasing Stored Energy

Solar cells capture directly the only net external source of energy arriving on Earth: light from the sun. But we will need to do more than capture. We will need to store this energy and we will insist on taking it with us when we move about. When the sun is down, we will want our clocks to keep running. When we require a burst of energy more intense than what the sun can provide—when we wish to go somewhere at seventy-five miles an hour on a cloudy day—we will require stored energy. We wish to transport energy across space and time.

We wish to transport energy across space and time.

The laws of physics insist that all we can do is convert energy from one form to another, and not create something from nothing. Energy harvesting is a translation among various formats—photonic, electronic, kinetic, thermal, chemical, gravitational—but is neither an act of net creation nor one of destruction. Only in nuclear power generation is there true alchemy: mass transmogrified into energy, $E = mc^2$. Fossil fuels—chemically stored energy—embody the same principle of energy conservation. Once-living vegetation harvested the sun's power through photosynthesis, energy later converted into oil, coal, or natural gas. Fossil fuels are stored solar energy preserved in chemical form. What, then, is the problem with solar energy stored in rotten vegetables? The issue is not the method of production but the by-products such as carbon

dioxide, the greenhouse gas, released when its energy is converted between forms.

What is so compelling about fossil fuels? The fuel we burn in our cars has some remarkable properties. Judged in energy per mass, or in energy per volume, it is compactly powerful, hence portable. The measure of energy per mass is important to us: we don't wish to weigh our cars down with many tons of fuel; once we've filled the tank we are glad that the car's weight is only modestly increased. Energy per volume is important too: fuel's volumetric efficiency ensures that our tanks are small compared to the total size of our cars.

This is not to defend the use of fossil fuels, but instead to identify what keeps us addicted to them as means of portable stored energy. Once we identify the ingredients that contribute to our gasoline fix, we may be able to break the cycle of dependence. The qualities of volume- and mass-efficiency help us understand in part why we have yet to budge from the overwhelming use of gasoline. They identify a clear and quantitative benchmark for comparison with clean energy-storage alternatives.

In addition to pollution during energy release, there is a further strike against fossil fuels. There will soon come a time when our extraction of these fuels from the earth reaches a peak and their cost rises further as a consequence. As this happens, we will increasingly feel the urgent need to capture and store energy on our own rather than rely on nature's fossil fuel legacy. Until now we have been living on borrowed energy.

Hydrogen fuel cells and batteries both provide means of storing energy. Hydrogen fuel cells consume hydrogen fuel, along with oxygen taken in from the air. They produce electrical energy and, as

a by-product, water—thus clean emissions. The fuel cells release stored energy—energy that had to come from somewhere. If the hydrogen fuel was produced with energy captured from the sun's rays, the cell is solar clean; if it was produced by burning coal, it is only as clean as coal.

Fuel cells are sandwiches, with electrical contacts on top and bottom, and in the middle a material designed to allow hydrogen atoms to flow. Hydrogen is fed in from one side of the sandwich, oxygen from the other side. The hydrogen atoms are separated into protons—the positively charged nuclei of the hydrogen atom—and electrons. The protons migrate through the middle of the sandwich to reach the other side, where they meet up with the oxygen, producing water and heat. The electrons reach the other side of the sandwich by passing through a separate route, one in which an electrical circuit harvests their energy. The meat of the sandwich is crucially important in its performance in transporting protons. By optimizing the length of the twigs emanating from the trunk of the molecules that make up this polymer, researchers have engineered the material to conduct protons efficiently. Chemical engineers have reinforced the trunk of the polymer, conferring mechanical, chemical, and thermal stability.

Fuel cells also require a catalyst, a chemical that brokers the separation of hydrogen fuel into a separate proton and electron. Platinum gives the best performance, but it is expensive. Recently, researchers have built nanoparticles of platinum to maximize the surface area of exposed catalyst for each dollar's worth of platinum. This nanometer engineering has proven so effective that the amount of platinum has been reduced by a factor of four without compromise to performance and lifetime. Fuel cells today are

60% electrically efficient. Because they emit pure clean water alone, they are attractive not only to power cars, but also to commercial buildings, homes, and portable devices. Imagine rather than recharging your laptop battery, refueling its hydrogen storage card.

Fuel cells to store and release electrical power cleanly are now usefully efficient. They are compact and lightweight. Their byproducts are innocent. Why are we not all driving hydrogen-powered vehicles and making espresso from their emissions?

Generating Hydrogen Fuel

Hydrogen is the most abundant element on Earth, but the molecular hydrogen gas used in hydrogen fuel cells is hard to get a grip on. It is so light that it will rapidly shoot up into the air, displaced by heavier gases such as oxygen and nitrogen. Thus while hydrogen is abundant, it is not readily available in the form we need for use in fuel cells. We are surrounded by it, but it is embedded in water or in hydrocarbons. Water is a stable, low-energy molecule: the product, not the ingredient, in the fuel cell reaction. The fuel to be consumed, and from which energy will be released, must be more energetic, more reactive than the product.

How do we produce the needed hydrogen fuel? A simple, efficient method involves splitting water using electrical energy, producing hydrogen gas. This is energy efficient, converting electricity into stored chemical energy. If the electrical power comes from the power grid, then splitting water is as safe or unsafe, as clean or dirty as the method used to power the process: hydroelectric, nuclear, coal, wind, or solar.

Michael Grätzel has recently demonstrated the direct conversion of solar energy into hydrogen for use in fuel cells. Even without building a new device, Grätzel could have used one of his efficient solar cells to produce electricity with 10% efficiency, and with it power water-splitting to produce hydrogen. The system would have an efficiency of about 7%. Instead, Grätzel used light to produce hydrogen directly, and in one stage rather than in a sequence of separate steps. (Risotto, similarly, would cost less time and energy if only one could sauté the shallots and toast the arborio rice in a single coordinated action.) Grätzel harnessed the energy contained in two differently colored particles of light to cleave water. Because cleaving one water molecule required more energy than is available within a single visible photon, Grätzel connected two light-harvesting systems. The first part absorbed blue light from the sun; the second used the green and red portions. The sum of the energies in these two portions of the system, and of the sun's spectrum, enabled them to produce hydrogen.

Grätzel harnessed the energy contained in two differently colored particles of light to cleave water.

In this way Grätzel produced molecular hydrogen suitable for use in fuel cells by directly splitting the water molecule. All of the energy came straight from the sun. In principle this method could result in an efficient, self-contained device. Grätzel's device converted sunlight to chemical energy with 4.5% efficiency—promising, but not yet approaching the tens of percent possible. Grätzel's two-photon approach was not entirely without precedent: his solar water-splitting procedure mimics plants' natural photosynthetic process.

Transporting Energy over Space and Time

Now that we have hydrogen, what do we do with it? Fossil fuels have raised expectations for storage: it's easy and comparatively safe to refuel one's car, and gasoline packs a large amount of energy per volume and per mass so that the tank is neither bigger nor heavier than the car. Pure molecular hydrogen, fortunately, is efficient in its energy content per mass: whereas a modern car burns 50 pounds of gasoline to go 250 miles, only 10 pounds of hydrogen would be required to go the same distance. It is hydrogen's tendency towards expansiveness that disadvantages it. Were it stored at atmospheric pressure and room temperature, these 10 pounds would fill a balloon 15 feet in diameter. Unfortunately the ice-cream truck and dirigible markets are too specialized to capture GM's attention.

We can compress hydrogen, squeezing it into a smaller volume with the aid of pressure. We can confine our 10 pounds of hydrogen, for example, to a simple, cheap steel tank five times larger than today's automobile gas tanks. High-pressure tanks are now being made using carbon-fiber-reinforced composites, halving the volume needed. Even higher pressure tanks can be made, confining more fuel to a given volume, but they do pose certain challenges. Putting a gas under tremendous pressure is dangerous, creating the possibility of explosion, and in Japan vessels with such high-pressure tanks are barred from the roads.

Gasoline is stored as a liquid, so why not do the same for hydrogen? This can be done, and it's extremely efficient in the use of space, but the tank needs to be kept at minus 423° Fahrenheit. The same liquid hydrogen launches the Space Shuttle, though comparison with its safety record and price-per-vehicle might raise concerns among automotive executives. BMW has nonetheless built an automated liquid-hydrogen filling station and has demonstrated cars running on hydrogen. Loss due to evaporation must be managed: to date it has been brought down to about 1% of the hydrogen's mass each day. Long-term airport parking looks less attractive still under these circumstances.

Nanotechnology may provide one interesting solution to the hydrogen storage problem. With the right chemistry, temperature, and pressure, hydrogen can be made to deposit itself over surfaces. Nanoporous materials have a maximum of surface area per volume, just like the villi of the intestine. However, since the forces that will stick hydrogen to a storage surface work only for the first molecular layer, only one single layer of hydrogen molecules can be accommodated. Huge surface areas contained in small volumes are thus required, and nanotextured materials offer a solution.

Theory says that, in the best case, four hydrogen atoms can be stuck on to a flat carbon sheet containing ten atoms. Such a storage system would be 3% hydrogen fuel by mass, the rest excess mass associated with the carbon sheets. At practical temperatures, 2% fuel by mass may be stored. Curving these sheets to make tightly wound nanotubes enhances the forces attracting hydrogen molecules, giving a 25% increase in hydrogen stored. Yet another argument in favor of voluptuousness: nanozaftig energy storage. Whether in fact hydrogen storage has been made efficient in the lab using nanotubes is less clear. The most striking result, a claim of 6 to 8 mass% reversible storage, has not been reproduced. Storage at about 1 mass% has been achieved.

Efficient hydrogen storage remains a challenge. In addition to dealing with technological questions, the hydrogen fuel cell industry will need to face the question of consumer acceptance. Everybody who has heard of the *Hindenburg* catastrophe associates hydrogen with fiery death. Hydrogen fuel cell advocates point out that hydrogen is nontoxic and so weightless as to dissipate in an instant if released. Whereas a Ford Pinto running on gasoline bursts into long-lived flames, one running on hydrogen spews a brief fiery plume high into the sky. Whatever the rational merits of these arguments, the manufacturers of automobiles and consumer products that are fueled by stored hydrogen will need to persuade their customers—intellectually and emotively—to trust the new technology. Hydrogen must permeate our hearts and minds before we will consent to its penetrating our proton exchange membranes.

Energizing the Future

In the energy sector a vast chasm of opportunity exists between the limits of physical possibility and the realities of present-day technology. Certain fundamentals must be obeyed, energy conservation first among them. Its main consequence is that we cannot continue to rely on non-renewable resources. We need to become active in capturing energy ourselves. There is no fundamental obstacle to harnessing the sun's power efficiently, no reason why we cannot do so cheaply and flexibly. There is no underlying barrier preventing us from storing and releasing energy with the concomitant release of clean steam instead of greenhouse gas.

If there are indeed no fundamental barriers to cleaning up our energy habits, then innovation offers the prospect of a solution. Nanotechnology has applied ideas from disparate fields—Michael Grätzel's solar cells, for example, take inspiration from high-surface-area textured biological structures and light-harvesting plants, and add insights from chemistry, physics, materials science, electrical engineering—to solve a technological challenge of huge importance to humanity's relationship with our natural environment.

5

Protect

Environmental enforcement agencies need to sniff out threats to our natural world at the earliest opportunity. They need to make prospective polluters believe that they will not get away with excess: place a speed trap at every intersection. Military and security agencies are equally concerned with sensing what is in the air we breathe and the water we drink. They need to warn of dangerous chemicals at the most minuscule levels, before these toxins threaten health. Such sensing needs to be done early in time and remotely in space, giving us the chance to avoid danger.

At the same time, industry needs to find ways of controlling and reducing the emissions it produces. This is one area in which nanotechnology has been playing a critical role since well before the word became popularized. Synthetic zeolites—materials filled with identical, nanometer-size pores—have been used and improved in petroleum refining to make lead-free gasoline; in catalytic converters to reduce our cars' noxious emissions; and in the production of phosphate-free laundry detergents. New refrigerants to replace ozone-depleting chlorofluorocarbons also rely on zeolites.

Nanotechnologists are using nanoparticles to bind heavy metals such as lead and cadmium and sequester these dangerous materials out of the groundwater supply. The hazardous waste from the Three Mile Island nuclear power plant was in part remediated using zeolites. Trying to undo the effects of our past wrongs—tackling the challenge of environmental recovery—is an emerging area for nanotechnology, one seeing significant investment from the U.S. Environmental Protection Agency. And while nanotechnology is being used to care for the environment, it is also being investigated for the risks it poses, such as if nanoparticles were to be released into our streams and rivers or into the air. Examining the environmental impacts of nanotechnology and devising an appropriate regulatory strategy is crucial.

Trying to undo the effects of our past wrongs is an emerging area for nanotechnology.

Sensing Threats to the Environment

If our own sense of smell were as acute as that of a dog, a child could smell its mother in another room, the border patrol would know when drugs were being smuggled in the car passing by, and a soldier clearing a field would know where the landmines were. Dogs' olfactory finesse proves that detection of molecules at minuscule levels of concentration is possible. Environmental monitoring would benefit tremendously from being this sensitive and discriminating, enabling rapid and decisive response to such dangers as the toxic waste from the factory upstream or nerve gas from terrorists.

Today's most sensitive chemical detectors are unwieldy, powered by batteries and weighing many pounds. These sensors could be made more convenient—even wearable. What if each of us could contribute to providing early warning of chemical and biological agents, or more routinely sense the extent of pollution in our local environment? We would thus gain a maximum of opportunity to protect ourselves from these threats.

Seacoast Science in Carlsbad, near San Diego, is one company making accurate, lightweight sensors. Its prototype is designed to be clipped into the shirts of U.S. Army soldiers. Each of Seacoast's sensors consists of a plastic layer that lies between two electrical contacts. The chemical to be detected enters the sensor and is absorbed in the active layer. This alters the electronic properties of the device, a change that is readily detected electrically. Each of these sensors is well under a millimeter in diameter.

The technique relies not on a single universal sensor, but on an array of ten, each one sensitive to a distinct class of chemical. Various gases change these polymers at different rates. For each gas to be analyzed, a ten-sensor signature is measured, and the gases are distinguished by their fingerprints, the relative strengths of their awakening of the different sensors. While such a chip is not so universal as to distinguish any gas from all others, it does allow more than ten different gases to be analyzed correctly.

Tim Swager and his team at MIT have taken a different approach to the sensitive measurement of specific neurotoxins. They do not measure changes in electrical properties, but instead look for changes in how light is produced in their chemically sensitive polymers. They work with polymers that produce light with vastly different efficiencies depending on whether a particular gas molecule has bound to its engineered receptor molecules.

Swager demonstrated what he calls molecular amplification, an approach that increases the change in the sensor's state in the presence of the neurotoxin to be analyzed. When even a single toxin binds to his sensor, the toxin prevents light emission along the full extent of a polymer chain. Because of this molecular amplification, simple light sources, detectors, and circuits can be used to detect the lowering of light-production efficiency. This makes for a strong—and potentially cheap—early signal, warning of the presence of toxins.

Swager has sensitively detected the explosive TNT, important in applications ranging from airport security to clearing landmines. TNT sensors are already available in airports, but they are expensive and not portable. Dogs work well, but they get bored, and their considerable talents could be better used elsewhere. In his TNT detector work, Swager solved a problem that had plagued previous work. In past detectors, polymer molecules had lined up into organized piles like matchsticks in a box. They filled up space efficiently and reduced the diffusion of gas into the detector. The change in brightness of the detector in the presence of TNT was thus reduced. Swager designed new polymers made from molecules that are not flat, and thus do not stack, keeping their brightness high when no TNT was present. When exposed to TNT, the detectors showed a strong decrease in fluorescence intensity.

The researchers also showed that other molecules, structurally similar to TNT, did not provoke a response in their sensor: their device was discriminating, even between TNT and similar molecules. The device was also astonishingly sensitive: whereas sniffer dogs achieve few parts per trillion sensitivity, Swager's sensors demonstrated one hundred parts per quadrillion. The detector's response was rapid and reversible, allowing the same sensor to be reused.

Swager's team has also applied his methods to detecting nerve gas. The molecules comprising nerve gas inhibit the action of a chemical within us that is critical to nerve function. Normally, signals pass through the nervous system with the aid of a neurotransmitter. To recover and be ready to transmit the next impulse, this transmitter must rapidly be broken down so that we can respond to the next stimulus. The fastest-acting enzyme in our bodies is in charge of this, breaking down the neurotransmitter in eighty-millionths of a second. If this enzyme is blocked by nerve gas, we become paralyzed. Swager and his group made sensors that respond specifically to the way in which nerve gas interferes with these critical enzymes. To read whether their device was reacting to nerve gas, the researchers detected the production of light as in their TNT detector. And once again they successfully sniffed, with great sensitivity, their target agent.

Thanks to the nanometer design of the molecules used to make them, threat detectors are now highly specific, sensitive, convenient, and portable. The next step will be to enable smart-suits that take action in response. Protective fibers and fabrics integrated into a smart battle-suit will automatically neutralize bacterial and chemical attack agents such as nerve gas. Materials will use nanopores that close up upon detection of biological agents, or neutralize chemical toxins. These smart materials are among the topics being explored at MIT's U.S. Army–funded Institute for Soldier Nanotechnology (ISN).

Combining and Relaying Sensory Information

Sensing threats in one place is a great start, but what we really need is to combine information accumulated across a battlefield, a chemical waste site, a community, or a nation. We then need to network the sensors and transport the information back to a central point for analysis and, ultimately, for action. A new strategy has recently emerged. Motes, co-developed by University of California at Berkeley and Intel Corporation, are miniature, self-contained battery-powered computers that use wireless links to exchange information. Also known as smart dust, motes will be designed to be so small, cheap, and connected that thousands of millimeter-size computers could be thrown like pixie dust into a field, an airport, or a sensitive environmental area, forming a completely connected network of sensors.

Historically, telephone networks and computer networks have been set up systematically, with addresses assigned to each member of the network so as to ensure orderly communications. However, planning such networks out in advance is not possible when we desire spontaneity. Instead, the network changes in time—as we tear off our networked chemical protection suit, put on a glove that has its own sense of touch, or connect a camera to our artificial retina. In such cases we need networks that organize themselves without outside intervention. This is what motes do: they sense one another's presence and form their own ad hoc networks. Engineers have developed an operating system known as TinyOS to allow

> Motes are miniature, self-contained battery-powered computers that use wireless links to exchange information.

programs to run over their motes. They have already used the motes to manage farms, monitor structures for earthquake damage, control industrial manufacturing, and inform troops in the battlefield. In the future they will help firefighters and rescue workers to communicate in dangerous environments. They may lead to interactive games in which we can share full sensory experiences, including sight, hearing, smell, taste, and touch: boxing at a distance.

Environmental monitoring by motes has been shown on Great Duck Island in Maine. A team deployed a wireless sensor network to monitor microclimates around the burrows used by a rare bird. The group monitored the bird's habitat without disrupting it. The motes were made of miniature computers, batteries, memory, and wireless transmitter-receivers. Built into them were sensors for temperature, humidity, pressure, and light. The motes to this day continue to measure these quantities and relay their findings, with the data hopping, from mote to mote, back to a computer base station. The information is fed into a satellite link that gives the researchers real-time access to the data over the Internet wherever they are.

Filtering Out Harmful Molecules for Environmental Protection

Measuring pollutants and toxins before they threaten our local environment is an excellent beginning, but we also need to prevent such emissions in the first place. Zeolites are solid materials dotted with nanometer-size pores that have for decades been making chemical processes more efficient and less toxic. Each of us has zeolites, for example, in our car's catalytic converter and air conditioner.

A large fraction of the space inside a zeolite is empty. The pores are the size of molecules, about 1 nanometer in diameter. Zeolites, which are crystals consisting of regular repeating elements, allow the pores to be made of identical sizes. Zeolites can be made with a variety of chemical compositions, allowing further tailoring of their functions. Because they have a gnarled inner surface, like the vast area in the pores of a sponge, most of a zeolite's atoms lie at or near a surface, and zeolites are therefore used to promote and control chemical reactions using these surfaces. They also function as molecular sieves, their small pores allowing small molecules to enter while excluding large molecules.

Molecular separation and purification take advantage of this sieving property. Shape-selective catalysis relies on zeolites to admit only the smaller reactant molecules, or to release only the smaller product molecules, or to control a reaction by ensuring that the intermediate products along its pathway to completion are limited in size to the pore diameter. The chemistry of zeolites' composition provides a refined level of control over their behavior. Those rich in silicon dioxide, the molecular building block of glass, shun water. They can be used to remove potentially cancer-causing organic compounds from water, letting the purified water pass through. Those with less silicon dioxide have the opposite effect, allowing undesired water to be removed from the organic solvents used in so much of chemistry and needed in their purest state.

Catalytic converters use zeolites to reduce polluting emissions from tailpipes of automobiles. They destroy 97% of hydrocarbons, 96% of carbon monoxide, and 90% of nitrous oxides produced by car engines, removing these materials before they reach the atmosphere. Today they are key components in every car driven in North America and are required in 80% of cars around the world. Catalysts based on zeolites are also widely exploited in petroleum refineries, where crude oil is separated into gasoline, jet fuel, heating oil, lubricating oil, and asphalt. One barrel of crude petroleum contains only 30 to 40% gasoline, but our needs demand more than 50% gasoline, so that separation is not enough. To meet demand, "cracking"—breaking down of large molecules such as those useful in heating oil, turning them into gasoline—is needed. Synthetic zeolites have been used in cracking since the 1960s.

A further refinement to the grade of fuels—high-octane gasoline—is required in high-performance cars. Much of what comes out of crude oil, even after cracking, does not have enough octane to burn well in these engines. At one time, lead was added to increase fuel's octane level, but when legislation mandated less lead in gasoline, new methods had to be found to produce higher-octane fuels. Zeolites are now used to transform larger molecules in crude into high-performance fuel with more octane. Octane levels can be further increased using a zeolite to separate out the lower-octane component.

Zeolites are used in natural gas as well. After it is extracted from the ground, unprocessed natural gas contains moisture, nitrogen, and carbon dioxide, to name a few of its contaminants. These need to be removed so that pipelines don't freeze or become corroded. Zeolites have been designed to trap carbon dioxide and water molecules but allow methane, the useful component of natural gas, to pass through. Zeolites are also used to remove 99% of ozone-forming, smog-causing nitrogen oxides from the emissions coming from natural-gas- and diesel-burning electrical power plants.

Molecular sieves have also been used in treating radioactive wastes. Zeolites helped clean up Three Mile Island and other hazardous waste sites. In fact, environmental remediation may be among the greatest applications for zeolites. Robert S. Bowman, a hydrologist at New Mexico Institute of Mining and Technology, has modified zeolites to make them treat a wide range of contaminants but let purified water through. These could in future be used to intercept contamination spreading through groundwater. Part of Bowman's contribution has been to show that the large quantities of zeolites needed in such applications can be made with the right particle size and at low cost. Sarah Larsen at the University of Iowa has developed nanometer-size zeolites that can suck volatile organic compounds—chemicals in paint and many cleaning fluids, and shown to cause cancer in animals—from the air. She has also used these compounds, with the addition of light from the sun, to trap heavy metals for removal from the environment. And at the University of California at Riverside, Wilfred Chen has made nanoscale materials that seek to reduce concentrations of dangerous heavy metals in water. Chen is engineering proteins that are highly selective to particular metals with the goal that they can selectively remove cadmium, mercury, and arsenic from the environment.

Quantifying Nanoparticle Toxicity

Nanoparticles are good at environmental detection, protection, and remediation largely because they are so chemically reactive. This fact, then, raises the question of whether nanoparticles could be harmful to the environment, and whether, if they seep into our waterways and the atmosphere, they could be dangerous to humans. Vicky Colvin at Rice University, in Houston, is looking into the toxicity of nanomaterials, focusing in particular on nanoparticles. She recently found that the toxicity of buckyballs depends entirely on the chemical state of the nanoparticles' surfaces. Colvin compared toxicity in nanoparticles whose surfaces had been chemically altered to make them soluble in water with those whose surfaces were unmodified. The unmodified nanoparticles are not in principle soluble in water, but they can nonetheless be present at low levels in the parts per million. When these untreated nanoparticles exist in water, they tend to cluster and stay afloat as long as they are few in number.

As Colvin points out, toxicity can be good or bad depending on context. Nanoparticles that kill cells need to be regulated lest the particles seep into the environment, remain active for long periods of time, accumulate, and unintentionally get ingested by humans. On the other hand, if toxicity can be controlled it can be harnessed: researchers designing new therapies for cancer hunt for materials that can selectively kill malignant cells, and others are always searching for better ways to control bacteria. Colvin found that, when she made subtle alterations to the buckyballs' surface, she could change their toxicity to cells by a factor of 10 million. When Colvin did not treat the surface to help buckyballs dissolve in water they clumped up and produced oxygen that attacked cells' membranes. When she did treat the surface, the buckyballs didn't generate the reactive oxygen, and their toxicity was minimal.

Researchers designing new therapies for cancer hunt for materials that can selectively kill malignant cells.

Researchers have also looked into the effects nanoparticles have on living animals. The first verified studies of single-walled carbon nanotube toxicity appeared in 2003. Two research groups found independently that granulomas—inflammations typically associated with the response of the immune system—developed in rats and mice when nanotubes were introduced into the animals' lungs.

This is different from what happens when larger nanoparticles made from the same materials are introduced. One unanswered question in these studies is whether it was not the nanotubes, but instead a metal used to make them, that brought about the granulomas. Researchers also debate whether the levels of exposure were practically relevant—would a subject ever inhale such concentrations of nanotubes into his lungs.

Colvin notes that the young but extremely important area of nanoparticle toxicity is in heated debate. Skeptics and proponents of nanotechnology have each made sweeping claims. Colvin argues that if nanotechnologists were to work immediately with toxicologists to bring technical data to the debate, the actual risks would become better defined, and such data would allow the study of environmental consequences before the nanotechnology and its products become pervasive.

Could nanomaterials have distinct, and not necessarily easily anticipated, environmental and health effects compared to the atoms of which they are constituted? Many would say yes. When we engineer nanometer-lengthscale materials to have properties that are distinct from either their atomic constituents or their larger bulk forms, their toxicology is likely to differ as well. Those nanoparticles that have the potential to be released into the environment must be studied for their environmental effects; and those that humans can come into contact with must be studied for their toxicity. Precisely this has been done in Colvin's and others' early studies. Superficially similar materials, differently chemically modified and taken up, do indeed produce vastly different outcomes that require systematic analysis.

As a second question, we should ask, Should the ways we deal with possible nanoparticle toxicity be different from the ways we deal with toxic substances in general? A regulatory framework that worked well for suspected environmental and human toxins could successfully be applied to nanoparticles too. Nanotechnologists need to recognize that their new materials have the potential for harm as well as good, and should begin looking systematically at toxicity during the development of new materials. The environmental domain illustrates starkly the combined opportunities and threats of any powerful technology: what can be used for good can have deleterious effects as well. Nanotechnologists must thrust ourselves into vigorous public discussion not only over how we can derive the greatest benefits from our innovations, but also over how we can guard against the potential downsides of our newly created materials.

6

Emulate

Nature is an ingenious nanoscale designer. She builds the shell of abalone by laying down layer upon layer of the crystal calcium carbonate. The butter to make her phyllo pastry cohere is an organic glue. While it makes up only a fraction of the mass of the shell, the glue, combined with the laminate structure, makes abalone three thousand times more resistant to fracture than a calcium carbonate crystal of the same thickness. Spider silk is another example of Nature's design: when scaled for fair comparison, the energy needed to break it is one hundred times greater than for steel. The molecules that whirr within us are magical, miniature motors with astonishing performance. The nanometer system that burns the fuel stored in sugars, producing movement within cells, can run at 800 nanometers per second against considerable forces: scaled to the size of an automobile, this molecular motor would travel as fast as a car powered by a jet engine, readily breaking the sound barrier.

In biomimetics, science imitates life imitates art: researchers learn from Nature, borrowing from the genius of her methods. They draw inspiration from her masterly materials and their underlying architectures, endowing matter with striking new properties. Biomimeticians have even co-opted Nature's workers, coercing genetically engineered viruses into making nanotextured materials of our choosing but of Nature's manufacture.

Building Materials Using Viruses

The series of breakthroughs made by Angela Belcher—previously at the University of California at Santa Barbara and then the University of Texas at Austin, now at MIT—highlight the principles and process of biomimetics research. First learn Nature's methods, then persuade her to manufacture new materials to order, organized from the bottom up.

In 1996, Belcher studied how Nature builds shell of the abalone, the strikingly sturdy structural material made from alternating crystal layers glued together: plywood on the nanoscale. The layers that make up abalone are not identical, but rather alternate between two types of crystal. Both are calcium carbonate, but they differ in how their rows of atoms are arrayed—same contents, different pattern, like packing oranges into squares versus hexagons. Two types of crystals are strong along different directions of stress. Built from one crystal alone, abalone would break under modest pressure applied along its axes of vulnerability. Alternating types of crystals make abalone strong enough to protect it from attack from any angle.

First learn Nature's methods, then persuade her to manufacture new materials to order, organized from the bottom up.

Belcher wanted to know how Nature built according to this brilliant architectural plan. In the course of her investigations, she found that abalone alternates between the use of two different proteins to promote the growth of the sequential layers. The mollusk produces these proteins, one and then the other, to guide the spatial rhythm of shell structure. Belcher studied how the proteins worked, and found that each specialized in one single, specific chemical reaction. Each directs the growth of a particular crystal type and sets its orientation. These proteins, programmed into the DNA of the abalone, are the project managers that direct construction, turning Nature's bold architectural vision into a hardy edifice.

Belcher looked further into why the shells are so strong. In particular, she investigated the glue that holds abalone's layers together. The glue consists of long molecules, polymers that are sticky at either end to fasten the layers together. To study the strength of these molecules, Belcher grabbed on to one end of the molecule and pulled, measuring its resistance. To do this she used an atomic force microscope with a probe nanometers in size whose pulling force could be controlled and whose movements could be precisely measured. Belcher stretched the adhesive one molecule at a time, finding that the polymer did not resist pulling in a smooth way, but stretched out in a stepwise manner. Each molecule, it turned out, was a knotted rope, and additional loops unfurled suddenly as she pulled harder, releasing tension before the rope broke and before attachment to the crystal layers was lost. Belcher's findings explained how abalone can be both rigid and robust, not inclined to shatter: the threads in the glue could stretch out considerably before breaking, and it would take a great deal of work to undo all of the knots.

Belcher invoked Sisyphus's legendary punishment. As soon as Sisyphus pushed his rock to near the top of the hill, it rolled back down, and his work was all for naught, dissipated in the form of heat. In the adhesive binding together the layers of abalone, the role of Sisyphus is played by the would-be breaker of the abalone shell: the majestic sea otter. He applies considerable force, and in so doing stretches only slightly the structure of the shell. Before the strands of gluey string attaching the laminate layers break, their loops unknot, relaxing the tension in the string. Sisyphus the Sea Otter's investment of energy turns to heat. He never quite makes it to the breaking point of the thread, but instead is frustrated as the string returns to its relaxed state, the shell still intact.

Nature had come upon this ingenious design not deliberately, but by accident. DNA contains the sequence of genetic information describing the designer proteins that build the shell. Nature, through fortuitous mutations, attempted countless variations, and evolution selected the DNA that produced the most sea-otter-resistant shells. Much of Belcher's subsequent work has been inspired by this systematic trial-and-error approach. Through biology, so many billions of trials can be staged, and the outcomes so conveniently screened in search of rare but valuable successes, that the strategy provides a powerful way to engineer new materials. Belcher, unwilling to wait through millions of years of evolution, would mimic Nature's evolutionary process of variation and survival-of-the-fittest selection in an afternoon in the laboratory.

Panning for Golden Proteins

Belcher chose an objective for her game of evolution: to get Nature to line up nanoparticles of pure, crystalline semiconductor, make them all point in the same direction. The electrical properties of semiconductor crystals depend on whether their crystalline axes are lined up. Belcher's work with abalone had shown that proteins had this capacity: could she co-opt biology's machinery, honed by accelerated evolution, and use it to grow improved semiconductors, the materials that make computers, CD players, and communications systems work?

Instead of waiting for Nature's occasional mutations, she would search systematically through billions of protein options in a single experiment. She obtained a library of strands of genetic material that encoded a huge number of proteins. And in her game of evolution, Belcher would use her own criteria, hunting not for proteins that conferred the ability to outrun enraged wildebeest or seduce fecund cavewomen, but for ones that enabled construction of semiconductor crystals of the type, shape, and size she desired.

Proteins begin as chains of molecules drawn from an alphabet of twenty amino acids. They are necklaces of any length, but made from a restricted subset of jewels: pearls, diamonds, cubic zirconium, and so on. The sequence of jewels along the necklace is written in our DNA, and the translation from DNA into protein chain is a literal mapping: sets of three molecules on DNA specify which jewel comes next along the protein necklace. Proteins' range of behavior is stunning, though, because they do not remain as long chains, but instead fold up, contorting themselves into shapes and structures that depend on the sequence of molecules from which they are built. Protein folding is origami on the nanometer scale, and Nature does the folding. The shapes that result are various:

Protein folding is origami on the nanometer scale, and Nature does the folding.

ribbons, napkins, tetrahedra, spheres. Irregular, complex, and functionally sophisticated, a protein's shape and chemistry make it fit, like a key, into particular locks. Proteins open doors to action within a cell.

Belcher obtained a library of viruses whose DNA was identical, with a single exception: programmed into the genetics of the viruses in this collection was one of two billion different possible sequences. These sequences were located in the genetic code of each virus such that they would program the production of the proteins that would be displayed on the virus's outer coat. In the massive swarm of nearly identical viruses, all that would distinguish one from the others was the unique set of trailing protein streamers. Belcher set the swarm loose on a piece of semiconductor crystal. She washed away those that didn't stick, saving only the viruses that, by virtue of their proteins, stuck strongly to the semiconductor: survival of the clingiest. Those that stuck were saved, ready for the next round.

Belcher, having gone from studying to mimicking Nature, proceeded to recruit its collaboration to create new materials. She began by obtaining a large, pure collection of the subspecies of viruses she had found that stuck to the semiconductor crystal of interest. No more trying out different keys: now she picked one and made millions of perfect copies. To make the copies she co-opted the virus factory that is an infected *E. coli* bacterium: it cloned, on command, the virus winners of the early rounds. And then, instead of seeing whether her virus would adhere to a semiconductor, Belcher grew that same semiconductor *on* the coat of the chosen virus.

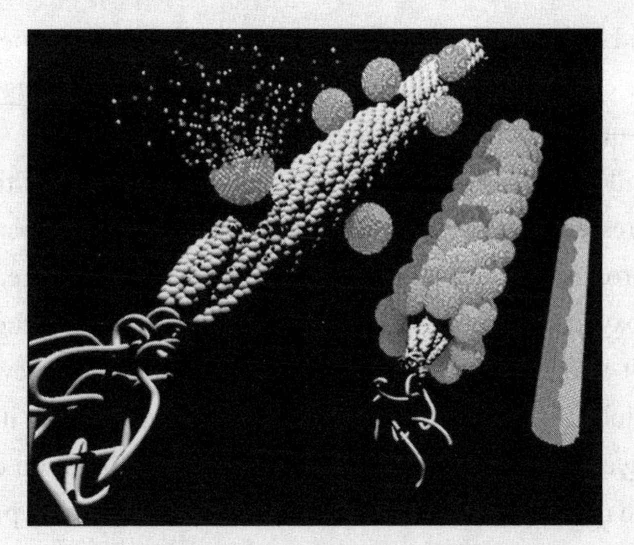

A virus enrobed in nanocrystals, their axes aligned by the orientation of the virus's protein coat. (Courtesy of *Science*)

The proteins on the outer coat of the viruses fostered growth of her nanoparticles. The proteins were not haphazardly arrayed on the virus, but instead lined up in regimented fashion along the organism's rod-like shape; they even organized themselves into longitudinal rows and circumferential rings along the cylinder: corduroy protein jackets. And likewise, the nanoparticles Belcher grew on the viruses lined up in lockstep with the ordered arrays of proteins. As she had intended, the achievement recalled Nature's growth of the abalone shell: the choice of proteins had facilitated growth of the crystal and defined the orientation of its atoms.

This technique, used by Belcher in the late 1990s to pan for proteins with specific properties, has also been used for well over a decade in developing new drugs. A protein on the surface of a cell that betrays a disease such as cancer allows researchers to hunt for protein keys that match the lock, which can then be used in targeted, seek-and-destroy chemotherapy. Personalized medicine in the extreme could recruit billions of viruses to develop a drug tailored to your unique condition: bespoke pharmaceuticals.

Not all imaginable applications are unambiguously for the good. Could researchers hunt for a molecule that would displace, even more strongly than existing gases used in chemical warfare, the life-giving oxygen attached to cytochrome a3? Released into the atmosphere, it would suffocate a city. Or could we speed up evolution? Hemoglobin is a brilliant transporter of oxygen, loading it up fast, carrying a great deal in a small volume, and releasing it on demand, but perhaps we could do even better: cultivate a superior protein—let's call it shemoglobin—that is lighter, more compact, and that loads, transports, and releases oxygen even more effectively. Would we use it to breed a nation of Lance Armstrongs?

If we can emulate the way biology builds matter and life, then the possibilities lie well beyond our current imagination. Instead of culturing pearls alone, why not seed, raise, and harvest a diversity of structural materials? Use selected genetic modification to grow the abalone-inspired carapace of a new, ultra-strong automobile. Squirt proto-glue protein between inner tube and patch, and command, "Tire, heal thyself!" Rather than manufacture layered photovoltaic devices based on spinach, instead plant, water, and grow them—then let them hook themselves into the electricity grid.

Powering Life through Molecular Motors

As in Angela Belcher's work, the world of biology is beautifully endowed with elegant structure. But Nature's objects are alive as well: life is movement. Angela Belcher's biomimetics has been used so far only to erect static edifices. Nanotechnologists are also seeking to put molecules into motion: turn the pose of molecules into a dance. And Nature has much to teach us in this regard. The graceful movements of the ballerina arise, ultimately, from the action of individual molecules. Sugar molecules in our bloodstream store energy, providing fuel. These dock to a motor the size of a molecule, changing its shape at the expense of energy stored in the sugar. Levers amplify a minor rotation into major motion: a deft flick of the molecular wrist. The motors are coupled to tracks made of long protein molecules, and amplification of motion transforms a shuffle into a series of bold *jetés* along the molecule's trajectory. In muscle contraction, the motors are tied down, fixed to one track, and move relative to the second one, thereby forcing the tracks to slide relative to one another. The structures that make up muscle can change their length by 10% in one-fiftieth of a second—running, flying, and playing the piano all depend on it.

Much like the astonishing powers it engenders in artists and athletes, motion on the nanoscale is a mesmerizing *Cirque du Soleil* performance. The characters in the cast are agile, flexible, coordinated contortionists. Kinesin is a molecular motor that organizes material in cells: he wears stilts seventy-two times his height and takes strides fifteen times longer than he is tall. Myosin powers the contraction of our muscles: she is a two-headed woman who walks head-over-head across a suspension bridge. These molecular gymnasts flaunt their prodigious abilities unabashedly: if kinesin molecules were ants, each one would carry a potato across the picnic blanket.

Molecular motors are crucial to every twitch within and among cells. It's also not surprising, given molecular motors' critical role in life from amoeba to tadpole to human, that motor breakdown has serious consequences. Many diseases are tied to flaws in the operation of one of the motors, often the result of a structural malformation—even the most subtle—in a particular protein. Disorders of pigmentation, loss of hearing or vision, kidney diseases, and neurodegenerative diseases are often traceable to motor defects. Sometimes motor malfunction can be lethal. Even invaders use motors: viruses succeed by co-opting the molecular machinery towards their self-serving ends.

The tracks on which molecular motors move come in a variety of types. Just as the gauge of a train's wheels must match the spacing of the rails, so too must the structure of the motor molecule match the features of its track. Some tracks are helices, twisting once every thirty-six nanometers. Others form tubes like water-park slides. Molecular motors crawl along the outer surface of the reinforced tube, a fat cylinder like the frame of a Cannondale bicycle.

Molecular motors are energy converters, transforming the chemical energy stored in sugar into mechanical action. The attachment of the fuel to a molecular motor changes the molecule's shape: a half-nanometer space opens up and the site where the sugar is bound rearranges. Though the motion is modest, it already involves one level of cause and effect, movement leading to rearrangement. At the beach, Laetitia Casta walks by and a man undergoes a conformational change: the optimal position of his Speedo is now altered, necessitating rearrangement. Importantly, the action—rearrangement—occurs now within the motor (protein molecule, Speedo), not simply the sugar (sugar, Casta). Movement is transferred to where it is needed.

Leverage must follow, since sub-nanometer reconfigurations will lead to slow progress along a track whose ties are spaced every

36 nanometers. A lever arm is fastened to the end motor molecule. In one motor, myosin, a rigid lever swings through an angle of up to 70 degrees, a movement known as the working stroke. Motors with longer lever arms take bigger steps and engender more action. (Further analogies with the Speedo effect need no explication.) A striking degree of amplification is the result: the stilt-walker's subtle rotation of the hip joints results in the stride of a giant.

More is needed, though, for a stride is not a stroll. Molecular motors either stick to their tracks and walk hundreds of paces, or hop off the track after a stride or two. Both classes are critical to life. Those that take many steps along their tracks are, for example, the great transporters of materials within the cell. These motors have two heads, ensuring that even mid-step at least one head is firmly attached to the track, and that even with a force pulling it backwards, the motor can plow forward.

Scientists recently looked into how these motors moved along their tracks. At least two possibilities were under serious consideration. Did motors waddle or did they climb? And if they climbed, were they inchworms or did they proceed hand over hand? Yale Goldman and colleagues at the University of Pennsylvania recently sought to answer the question by watching how a molecular motor walked along a track. They did so by tying a light-emitting molecular flashlight to its thigh. If it moved like an inchworm, the head would take a stride equal in length to the movement of the cargo. If it moved by hand-over-hand motion, the moving head would travel farther than the cargo in a single step. Goldman and his team showed that the head moves more than the cargo on each stride. Since Jeff Gelles at Brandeis University, in Massachusetts, had previously eliminated the waddle hypothesis, Goldman concluded in 2003 that this family of motors moves hand over hand. Goldman, in his subsequent plenary address at the Biophysical Society's annual meeting, gave his best imitation of myosin V's motion to the strains of James Brown's "Papa's Got a Brand New Bag."

Did motors waddle or did they climb? And if they climbed, were they inchworms or did they proceed hand over hand?

Another molecular motor protein induces cilia, the tiny hairs sticking out from the surfaces of cells, to flail about. Hundreds of nanometers in diameter, tens of micrometers long, cilia sweep dust and mucus from the lungs up and out. They sweep eggs along the oviduct, and their cousins, flagella, propel sperm. Cilia bend in directional waves, each motion like the cracking of a whip. First they lash out with a bold forward stroke. In the subsequent recovery phase, the cilium returns to its original position, gently unrolling itself. Net forward propulsion is the result. Cilia and flagella whip by bending the cores of their constituent microtubules using motor proteins. The tubules' structure turns sliding into bending, and power comes once again from sugar.

The true test of our understanding of how molecular motors work will be to design them to suit our purposes. Nature's ingenious machinery sets a high standard, but the payoff would be so great as to be well worth the challenge.

Inventing New Molecular Motors

Ross Kelly at Boston College recently made progress towards molecular motors of our own design. Appealingly, motors such as Kelly's can be made up of only a few dozen atoms, and a billion billion machines can be made in one batch. His work builds on past demonstrations of a molecular ratchet, a wheel hindered by a curved brake that favors rotation in one direction. Like a macroscopic ratchet, the molecule thus has a directional preference without which the motor would rotate in either direction with equal likelihood and not produce a net forward motion. Kelly was able to produce a 120° net movement of his molecular motor, but has not yet achieved continuous rotation.

L.E.J. Brouwer at the University of Amsterdam has made a molecular motor that functions like a piston. It is driven not by chemical fuel but directly by light. The light-induced power stroke turns optical energy into mechanical power. We can dream of muscles powered directly by the sun, avoiding the intermediary chemical conversion and storage.

In addition to ratchets and pistons, researchers recently designed and built molecular tweezers. Bernard Yurke at Bell Laboratories in New Jersey, with collaborators at Oxford University, showed that their engines implemented continuous operation—repeated motion. The researchers used DNA and took advantage of its property known as hybridization. The DNA molecule will stick together, forming a stable double-helix, only if its two strands form a complementary pair: that is, if the key matches the lock.

Yurke made his machine by mixing, in equal proportions, three different types of DNA molecules labeled A, B, and C. Strand A was complementary to the shorter strand B at one end, and complementary to C at the other. Thus the lengths of B and C add up nearly to that of A. Double-stranded, or hybridized, DNA is rigid, so the arms of the tweezers are stiff. Only the hinge is bendy. The tweezers lie open at this stage.

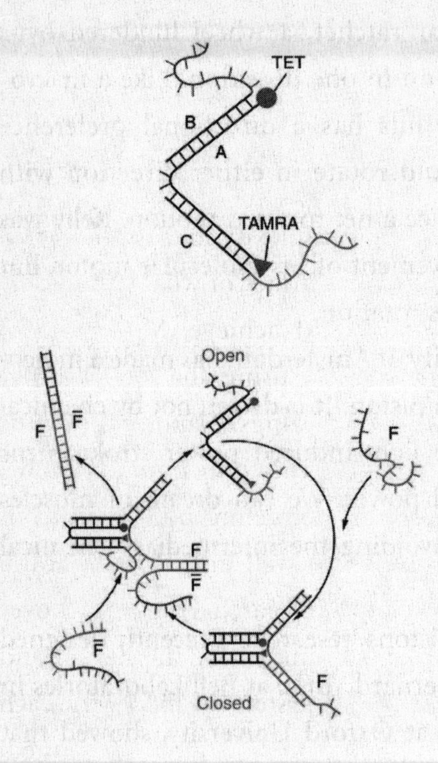

Yurke's molecular tweezers opened and closed through the addition of DNA. (Courtesy of *Nature*)

Introducing F, the fuel strand, led to the closing of the tweezers. F has a portion that matches the dangling ends of B and C, forcing the two arms of the tweezers to come together. Yurke's tweezers were reversible, since introducing the strand F̄, which matches up perfectly with F, steals F away from hybridizing with B and C. It cuts in within the molecular dance, stealing F away and allowing the tweezers to open back up. The product of the reaction—the spent fuel—is F F̄, a hybridized DNA molecule created after a cycle of tweezing is complete.

Yurke and his collaborators had managed to build a molecular motor through the simplest course of action: pouring beakers of molecules into one another. Brilliant molecular design, combined with simple actions, led to nanometer dexterity, with Yurke doing the thinking, but Nature the precise work.

Imitating Life

Life provides an inspiring model of complexity and sophisticated control. We are only now scratching the surface of what biologically inspired molecular manufacturing could achieve. Movement of Yurke's tweezers was invoked selectively, using the lock-and-key mechanism of DNA's complementary pairings. Nanotechnologists have now achieved control over molecules not only using top-down movement of atoms as with scanning tunneling microscopy, but also based on molecular design in which well-understood natural forces do the work. The kind of control researchers exercised over molecular motion suggests new opportunities for the design and fabrication of nanostructures. An entire system of machines, each with different sequence-specific behaviors, can be envisioned, operated by a series of lock-and-key fuels.

New reactions could resemble the cycles at work in our bodies, storing and consuming energy. Scaled up, they would mimic the layers and layers of development within us: from the molecular to the cellular to the realm of the sentient.

Information

Consider how far we've come in the information era, and also how we have but scratched the surface of possibility. The first stored-program computer, ENIAC (Electronic Numerical Integrator And Computer), was built in 1946. Based on vacuum tubes, it could add five thousand numbers in one second, or calculate the trajectory of an artillery shell in thirty seconds. ENIAC weighed 60,000 pounds and occupied 16,000 cubic feet of space. It also consumed more energy in calculating the trajectory of one shell than was needed to fire the shell.

Today's triumph of computer portability would have come as no surprise to early visionaries: a panel of experts predicted in 1949 that, one day, a computer as powerful as ENIAC might be as light, and consume as little power … as an automobile. In fact, today's cell phones are thousands of times more powerful than ENIAC and consume much less power than a single one of its 18,000 vacuum tubes.

The history of communications technology is as impressive. A single fiber-optic cable now carries 100,000 phone calls along glass finer than a human hair. One hundred years ago, a much bulkier copper wire could carry one such connection. The result is that we are continually connected—the sounds we make, the images we reflect, and the words we type can be sent around the world almost as if each of us was everywhere at once.

And yet there is so much more to be accomplished. What Ionesco-esque tales will we tell in 2050? "Grandpa, is it really true that you used to put large, radiating objects up near your brain to talk with your invisible friends?" "Grandma, did you really use your thumbs to punch out cursory, hundred-character messages using a three-by-four keypad?" "Do you expect me to believe your stories of large rigid folding windows carrying clumsy images of 'books'?"

Nanotechnologists are building matter from the molecule up to share and use information in new ways. Computers built from silicon transistors have been a triumph, but this strategy will eventually run out of steam, whether the end comes first for reasons of technology or money. Nanotechnologists are instead seeking to make thinking machines from molecules, imperfectly organized but programmed to work anyway. We are also working to connect people and machines, and people through *machines, not with clumsy keyboards but with wearable, flexible interfaces that suit human architecture. And we are evolving the Internet from today's mishmash of informatic currencies—photons and electrons clumsily exchanged with high commissions—into a seamlessly connected network powered by light.*

7

Compute

Today's electronic chip—the heart and mind of the computer—is made by mass-producing and connecting millions of identical tiny building blocks known as transistors. These devices do the simplest of things to the smallest units of information: add, subtract, and negate bits. Combined, these simple elements become powerful, so flexible that fast math, rich and lifelike graphics, and interactive games—all these become possible. The simple strategy of these building blocks, richly interconnected, enables a strikingly versatile machine.

Silicon, the physical foundation of chips, is an abundant element, predominating in sand, rocks, and windows. Pure, perfect crystals of silicon consist of ordered arrays of atoms. From these crystalline blocks of silicon are carved transistors, and in these devices electrical current is either allowed or forbidden to flow through a channel. They operate much as do locks controlling the flow of water in a canal. Instead of raising and lowering a physical barrier, transistors raise or lower an energetic barrier to the flow of electrons along the channel. Transistors' functions of flow and control share a common currency: the electron. Cascaded interconnection is possible as a result, the flow along one part of one canal now able to control that along another. This creates the potential for chaos, or alternatively, if operating under an inspired master plan, the possibility of computational complexity—an intelligent machine.

Each transistor must perform accurately, rapidly, and efficiently. Accuracy lies in the sharpness of the bit, or binary digit, that it represents: the depiction of a pure 1 or a pure 0 state without ambiguity. Rapidity is in the time the transistor can be switched from one state to another, today fractions of a billionth of a second. Efficiency comes from reducing the number of electrons needed to raise or lower a lock, analogous to minimizing the number of Wheaties the locksmen need to consume in order to acquire the energy to carry out each operation.

Printing Transistors onto Chips

To build a connected community of these transistors on a single platform is known as integration, and the result is an integrated circuit—a fully formed computer chip. This community is densely populated, but is without diversity: one inhabitant, the transistor, is cloned millions of times. Relentlessly reproduced, intelligently networked, simple transistors achieve a range of sophisticated function they would never achieve acting alone. The cloning procedure is critical to the success of this strategy.

One inhabitant, the transistor, is cloned millions of times.

Gordon Moore, founder of chip-maker Intel, showed legendary prescience in his 1965 article titled "Cramming More Components onto Integrated Circuits." Moore predicted that, as transistor cloning was further refined, the number of transistors on each integrated circuit would continue to double with each successive technology generation. The computing power of the chip would grow in proportion. The press dubbed it "Moore's Law" and the name stuck. When Moore made the prediction, there were thirty transistors on a chip; today there are a billion. The persistent and rapid growth that Moore observed continued unabated over four decades, with speed and sophistication doubling every eighteen months since the 1960s. Cheap, pervasive computers are the result, as is a semiconductor industry worth hundreds of billions of dollars.

Lithography is the means by which massive numbers of identical transistors are tattooed, all at once, in virgin silicon's atomically smooth backside. The technique also provides a means of tying transistors together to make circuits that make sense. Lithography is like screen printing using stencils, a way to copy images countless times. In one step a simple image is formed. A few overlays, precisely aligned, generate a complex polychromatic image, for example, a pink-and-white Hello Kitty sitting in the middle, her powder-blue friend Teddy at a distance.

Similarly, in optical lithography, a modest number of printing steps, aligning each image with the ones previously laid down, leads to the production of complex circuits. Screen printing allows the definition of a complex pattern—six whiskers, two eyes, zero mouths—in a single exposure. In optical lithography, details are printed onto the substrate as photons pass through openings in the stencil, specifying precisely and reproducibly millions of transistors. Moments later, the process is repeated on the next chip in the assembly line.

As in photography, a light-induced change in the exposed layer transforms a dance of light into a real physical imprint. The optical image is transferred onto the plane beneath, like crop circles formed when the sun blazes through round partings in the English permacloud. In optical lithography, the photosensitive material that translates light and shadow into tangible matter and its absence is known as photoresist. Once regions of the photoresist are exposed to light, a developer solution dissolves the polymer, which stands its ground if it has only ever seen darkness. An image in light, inverted, becomes a solid plastic film on the surface of the nascent chip. These photoresist stencils are then transferred into hills and valleys in the silicon landscape: deep trenches are etched when a reactive gas shovels away at the exposed semiconductor. To make the wires that link transistors together, snowflakes of aluminum flutter down onto the chip, and silvered hills rise up.

Snowflakes of aluminum flutter down onto the chip. and silvered hills rise up.

To impose on the light-sensitive polymer the pattern that will stencil subsequent shapes, vast flocks of photons launched from lamp or laser converge through lenses onto the surface of the chip. Only because these particles of light are so content to fly in formation, each minding its own business, can they form a perfect image on top of the wafer. Chargeless photons are indifferent to one another's presence. This property—the vast parallelism of light—is exploited *a fortiori* in optical lithography. If each independent pixel's worth of lithographic exposure were thought of as information to be communicated, today's lithographic systems would send 100 billion bits of infomation every second.

The planar lithographic process is central to Moore's Law of growth in chip density. Once a computer architect has invented a new chip design, new stencils are generated, and from them follow the shapes of penumbra detailing the corridors and crevices of the new, speedier, more sophisticated processor. The designer's circuit is ready for prime time.

Shrinking Transistors to Nanometers

Researchers have for decades warned of impending crisis, an imminent end to the Moorish era. Their argument is simple, fundamental, and correct. Light not only consists of particles—photons—but also is a wave, and this wave has a characteristic curvature described by its wavelength. If we lock in our choice of wavelength, then our ability to scribe transistors is constrained to a defined fraction of that wavelength. As Rebecca Henderson of MIT's Sloan School of Management pointed out in her 1995 work "Of Life Cycles Real and Imaginary: The Unexpectedly Long Old Age of Optical Lithography," engineers have been brilliantly successful in stretching the practical limits of lithographic patterning. Lithography has been improved continually to meet the needs of the next generation of chips. We have shortened the wavelength of light chosen for transmission through lithographic stencils, in the process taking photons' colors from blue to deeper blue and now well into the ultraviolet—the rays that burn our skin but which we can barely detect with our eyes. Lithography becomes more difficult, but not yet impossible, as we move to use photons containing larger and larger quanta of energy.

To progress, we will need to invent new lenses that can focus increasingly energetic, short-wavelength, ultraviolet light onto a chip. Spatial resolution has recently been doubled by taking advantage of light's wave character. Two same-sized ocean waves crashing into one another can, depending on how they line up in time and space, either reinforce or cancel one another. Their peaks can double in amplitude, and their valleys double in depth, creating a much sharper super-wave. Alternatively, peaks colliding with valleys achieve perfect mutual annihilation. This concept of wave interference is exploited in optical lithography to sharpen features on the chip: lightwaves' selective reinforcement leads to an enhanced exposure of photoresist, and controlled wave cancellation leaves features that are not to be exposed entirely in the dark.

With these technological refinements combined, lithographic systems are incredibly complex and costly: the systems are the size of small moving vans and have tremendous mechanical complexity. The assembled lenses are three feet tall and cost millions of dollars.

The assembled lenses are three feet tall and cost millions of dollars.

More radical avenues under development in the laboratory could postpone further Moore's last sigh. Extreme ultraviolet lithography would reduce by a factor of ten the wavelength of light currently projected through lithographic masks. This technique proposes using an exposure wavelength in the 11- to 13-nanometer range, ten times shorter than the wavelength of light used in today's most advanced experimental ultraviolet lithography systems. Materials, including glass lenses, relentlessly absorb these energetic ultraviolet photons. Extreme ultraviolet systems are thus designed not based on bending light traveling through lenses, but light reflecting off curved mirrors. The mirrors themselves are unconventional: traditional thin metal coatings will absorb too much ultraviolet light, so instead, complex, curved, layered stacks deflect light. In reality today these are imperfect, not as highly reflective as desired, and their number must be minimized so that the loss of precious light is mitigated. Finally, making an efficient, pure, and bright source of extreme ultraviolet-ray photons is itself the subject of research.

Beams of electrons can also be used to form patterns on silicon chips, but then the features must be written one at a time. Electron-beam lithography achieves nanometer resolution, a consequence of electrons' characteristic wavelengths of nanometers or less, much shorter than photons' hundreds of nanometers. The technique is already in widespread use today to form the original masks reused thousands of times over to print endless chips by optical lithography. This is the perfect use for electron-beam lithography: scanning a beam row by row to define each feature in sequence, it is ideal for making a single custom mask. As such, it is also well-suited to building experimental circuits in the research laboratory. This kind of lithography does not, however, benefit from the parallelism of optics ideal for mass-manufacturing identical circuits. Electron beam lithography transcribes using one monk with a quill; optical lithography would make Gutenberg proud.

Fixing Leaky Transistors: Calling for a Quantum Mechanic

The limits of lithography—even if they have turned out to be more forgiving, more susceptible to engineering improvement than originally believed—solve only one challenge to the perpetuation of Moore's Law. Improved lithography makes smaller transistors; but will smaller transistors always run faster and perform better? A single dimension, carved out using lithography, determines how long an electron takes to pass through a transistor. The gate length describes the extent of this critical reservoir separated by adjacent locks in the electron canal. The gate liberates or strangles the flow of current from the source; the electrons thereby emitted then trickle into the drain that collects them. The electrons can travel only so fast, at a speed limit established by silicon's semiconducting properties. Their velocity predetermined, electrons then traverse the channel in a time set by the gate length. Lithographers strive to control this critical dimension of gate length, ensuring rapid transit of electrons simply by shortening their commutes.

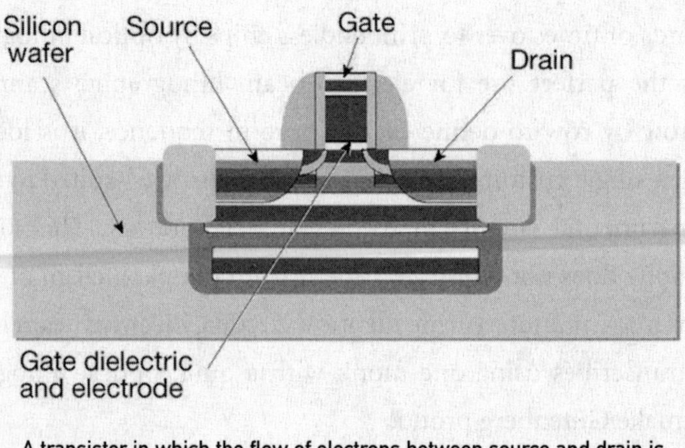

A transistor in which the flow of electrons between source and drain is controlled by the gate. Many such transistors are printed at once on a single chip and then connected to make a complex circuit. (Courtesy of *Nature*)

The vertical relationship between gate and channel is critical too. The genius of monolithic electronic integration—the ability to define millions of transistors on a single chip using a small number of steps—comes from the fact that lithography can be used to define distinct lateral features millions of times over in a single exposure. In contrast, the vertical sandwich of the device is the same everywhere and its uniformity is absolutely imperative. In the vertical direction, the gate and channel are separated by a thin piece of electrically insulating material, made by passing oxygen at high temperature over silicon to transform a thin layer of it into glass. As lithographically defined gate lengths are continually reduced, the thickness of this oxidized layer must be decreased, a phenomenon known as scaling. Transistors must maintain certain geometric proportions: a body–mass index (BMI) between 18.5 and 25 is normal; anything else imperils transistor health. Industry experts predict that by 2012 gates will need to be 35 nanometers long, and, for transistors to maintain properly scaled proportions, only 5 atoms tall. Fortunately, chip-makers such as Intel and IBM have gained control over the gate-making process, and they are already meeting this demanding requirement today.

Precise control over the chemistry of turning thin layers of silicon into sand allows this atom-perfect control. The devices that result, however, even when perfectly manufactured, are so thin as to experience electrons' quantum mechanical wave behavior. Quantum phenomena—guitar-string-like electron wave resonances—would represent an opportunity if electronic transistors were poised to take advantage of quantum effects; but relentless pursuit of Moore's Law has instead set them along a prescribed path such that when quantum mechanics comes into play, the performance of conventional transistors suffers. Electrons leak through ultra-thin gates via quantum mechanical tunneling because electron waves are everywhere—on both the gate side and

Electrons leak through ultra-thin gates via quantum mechanical tunneling because electron waves are everywhere.

channel side of the gate insulator. If one electron's tunneling induces further electrons to flow, a growing avalanche of electrons can go rumbling from gate into channel. This catastrophic leakage is opposite to what is desired, for the gate insulator is there to isolate gate and channel from one another. Imagine a canal in which the hydraulic controls raising or lowering locks start to leak, and even keeping water levels fixed requires energy. In already power-hungry, overheated electronic circuits, this is too high a price to pay.

Researchers have launched the quest for materials that allow thicker layers to separate gate and channel, preventing leakage while ensuring fast operation. These materials allow a different set of scaling rules to apply, as in the case of a person who is mostly made of muscle and in whom a higher-than-average BMI can be tolerated. (Brad Pitt, at 6 feet and 203 pounds, is officially overweight and 2 BMI points from obese.) Bulked-up insulating materials could further extend the persistent enforcement of Moore's Law. These materials, however, are not as easily formed as today's thin glass insulator gates, so readily put in place by passing oxygen over silicon. The capacity to transmogrify semiconducting silicon into insulating sand is among Nature's most precious gifts to the computer chip industry.

Restyling Computers Using Chunky, Piecey Nanotransistors

The engineers who have kept computer chips doubling in density every eighteen months have been surprising skeptics over the forty years that they have perpetuated Gordon Moore's early observation. By improving lithography, changing transistor geometry, and inventing more effective gate materials, researchers have stretched the capacity of integrated circuit engineering well beyond limits thought fundamental a decade ago. Even before we run headfirst into the hard castle wall of fundamental physical limits, wherever precisely these may lie, we may first become severely bogged down in the surrounding moat, bitten by piranhas. Making integrated circuits of sufficient speed and sophistication to meet our ever-growing appetite will first become too difficult and therefore too expensive. Even today, to build a chip-making plant costs $4 billion. The set of companies that can compete with the stakes so high is shrinking, and their reluctance to accept the investment risk is growing.

Even today, to build a chip-making plant costs $4 billion.

Nanotechnology effected not by carving from the top, but by planting seeds and growing from the bottom up, could turn chips on their heads. Researchers are seeking to persuade Nature to build computers through molecular self-organization. Avi Aviram, now at IBM at Yorktown Heights, and Mark Ratner, while both were at Northwestern University in Illinois, proposed in 1974 a diode—a device even more basic to electronics than the transistor—built not on semiconductor crystals but on a layer of individual molecules. In silicon, diodes are made by putting into contact two pieces of semiconductor that have different affinities for electrons. One material is luxuriously bathed in an abundance of these negatively charged particles; the other suffers from an extreme paucity. If electrical force obliges electrons to cross the junction between these devices, they produce a lush waterfall of current; anima and animus are joined, yin completes yang. In contrast, if an electrical voltage is applied in the opposite direction to make each side even more like itself, almost no charge can flow. Diodes thus respond differently to electrical forces of differing polarities, positive versus negative: they react enthusiastically to the carrot, but are unmoved by the stick.

The polar nature of the diode comes from its built-in asymmetry. With the two sides so dissimilar, there is a strong preference for one direction of flow over the other. A diode is a room half full of liberals on one side, half full of conservatives on the other. Propelling enemies together will ensure cacophony, the beating of breasts, the strangling of throats. If the groups are pulled apart, a cautious consensus will ensue. Because the room is so polarized, pushing the groups together has a strikingly different effect than keeping them apart: the room's asymmetry is manifested in the brouhaha or detente that follows.

Aviram and Ratner designed molecules that embodied this same asymmetry, calling the two ends of their molecule the donor and the acceptor. The donor was endowed with electrons; the acceptor craved them. Once such molecules were created in a liquid solution, they would be induced to line up on a solid surface, all with identical orientation. The asymmetry, now consistent over a vast field of molecules, would lead electrons to flow comfortably from donor to acceptor, but not the reverse.

Decades ago, this was an exercise of the imagination. Recently, some of the building blocks of molecular electronics have been built in the laboratory.

Cees Dekker and colleagues in The Netherlands demonstrated in 1997 a transistor in which the channel was formed by one molecule alone. Electrons flowed along a single carbon nanotube suspended across a chasm, forming a perilous rope bridge for electrons. In contrast with conventional transistors, but just as on a rope bridge, electrons were inescapably influenced by one another's presence. This fact led the single-molecule transistor to exhibit distinctive behaviors compared to conventional transistors. Because Dekker used a perfect single molecule, his devices, measured at the coldest possible temperatures, produced striking evidence of quantum mechanical behavior: abruptly bumpy relationships between current and voltage. The flow of current started suddenly, a consequence of the chunky nature of electrons, which come in ones, twos, and threes, and not half-sizes. A classical resistor would show smooth conduction.

Dekker's experiments were like playing a game of bowling across a saggy rope bridge. With no bowling balls in play, the bridge sags only mildly. If a bowling ball is launched rather tentatively across the chasm, the weight of the ball will induce further sag, and it will not make it across—no current will flow. If the bowling ball is launched with greater vigor, it may have enough energy to overcome the sag and make it to the other side—current flows. Moreover, if bowling balls are being launched at a rate such that more than one is on the bridge at a time, each one's further sag-inducing effects perturbs the experience of the others. Having more bowling balls on the bridge at one time impedes flow, and it does so in a way that depends on the exact quantized number—one, two, or three—of balls on the bridge. This is what Dekker found: that the number of previously launched electrons influenced the flow of the last electron added.

Dekker's experiments were like playing a game of bowling across a saggy rope bridge.

Dekker's findings implied that electrons were acting like coherent waves, extended in space along his cold nanowire. Over the

140-nanometer distance separating Dekker's contacts, electron waves spanned their entire chasm, feeling the effect of any other electrons on the channel. Dekker's quantum staircase relied on electrons sensing one another's presence rather than being dominated by the influence of local surroundings alone. Dekker found that exceedingly low temperatures were indeed necessary to see the quantum mechanical effects in electronic transport. At the lowest temperatures to which matter can be taken—minus 459° Fahrenheit—the steps of his quantum staircase were sharply defined. As the temperature was heated above the coldest physically possible to a balmy minus 458° Fahrenheit, the edges of the steps became more rounded, quantum behavior indiscernible against the smooth classical background: warm instrumental adaptations of Beatles music drowned out Tupac.

While appealingly based on Nature's manufacture of molecular building blocks, Dekker's approach illustrates another practical challenge to be overcome. Nanotube rope bridges need to be placed such that they span the chasm and are accessible to foot passengers on either side. In their first experiments, Dekker and colleagues sprinkled a mixed bag of nanotubes atop a chip containing thousands of proto-transistors—sets of contacts and gate electrodes missing only the channel itself. More often than not, zero, two, or three nanotubes fell between two contacts. Only occasionally did a single nanotube span the chasm to form a single-molecule device. Moreover, not all such nanotubes were of the right sort. When Nature forms these long molecules of carbon atoms rolled up to form a cylinder, she rolls in a variety of ways. She may join the two bottom corners together, or may roll at some angle that dramatically alters the electrical behavior of the carbon nanotube. Some of these nanotubes act like metals, indiscriminately conducting any electron in sight. Others act like semiconductors, selectively carrying only electrons of a certain energy. The bottom-up means of creating carbon nanotubes—an explosion inside a chamber that creates special soot—produces a mixture of the two types of nanotubes. To make a transistor using a single nanotube, however, only the semiconducting type will do.

Dekker's approach obviously does not mesh with the way we build electronic circuits today. Our chips rely on 100 million transistors working perfectly and programmably connected together. There is obviously a mismatch between top-down nanotechnology and the bottom-up technique, and an urgent need exists to link the two: we must start digging the Chunnel from each side of the channel such that we meet in the middle.

Programming Molecules?

In a branch of silicon electronics known as field-programmable gate arrays, the idea of first building mass-produced generic chips, and only later custom-programming each one, is in use. Could we do the same with devices made from molecules? Nature would build chips through stimulated self-organization, and, as with all she creates, each chip would then be physically unique, just as each flower in a class or each maple leaf on a tree shares genes but differs in the outward show of structure. Genetically identical twins are distinguishable, for in Nature, nurture combined with nature yields variability. After commissioning Nature to build complex but imperfect chips, molecular computer designers would then configure the processors—warts and all—for reproducible computation. This might be as simple as routing signals around the warts.

This strategy would need to rely on electrically programmable molecular diodes: designer molecules that, when an electrical voltage was applied, reconfigure such as to favor, or prevent, the flow of current. In one image, in the "up" position, rings would strangle off current, and in the "down" position, they would enable electron flow. Programming of the device would configure each ring into one state or the other.

Researchers have built molecular electronic devices on this principle, with the movements of organic molecules called rotaxanes used to turn on and off the flow of electrical current. The rotaxanes lie between two pairs of metal electrodes, which are required to make electrical contact. While necessary, such a structure made it hard to know exactly what was going on within the device. Critics have suggested that voltages intended to turn switchable molecules on and off may have changed the properties of the metal electrodes themselves. In the presence of an applied field, metal atoms in the electrodes could creep across the nanometer-size gap, migrating to forming filaments through which current could flow. These hypothesized "programmably leaky" devices might work, but not based on the elegant movement of rings on molecules. Beautiful diagrams of designer molecules would serve solely to separate the electrodes and decorate the press releases.

If and when such molecular devices and circuits are realized, how will they form the basis of a computer? In silicon chips, the millions of identical transistors are necessary, but on their own are insufficient. A moist pile of neurons may constitute a brain, but a mind is made up of connections. Transistors, similarly, must be connected to one another, and to the chip's communication system responsible for bringing in the information to be processed and carrying away the results. Computer scientists propose that molecular circuits' architecture will be different from that of silicon chips: we may be able to persuade Nature to construct arrays of molecular transistors, but if we decide to delegate to her the construction of our chips, we must not nano-manage her.

Stan Williams of HP labs, Hewlett-Packard's central research organization, has focused on building computers not using three-terminal transistors as in today's machines, but rather using Aviram and Ratner's simpler molecular diodes. Researchers would program computer chips made from these devices using software routines that ensure that the chip only uses those molecular devices that worked well, overcoming in this way the molecule-to-molecule variability of these chips. These arrays of molecules would still need to be wired together to allow programming and read-out of the state of each individual molecule. Researchers at the California NanoSystems Institute (CNSI) have shown one way to deal with the wiring problem: instead of defining nanometer-size wires sequentially using an electron beam, the CNSI team built a nanotemplate using a bottom-up approach. They grew a multilayered nano-sandwich, the slices of bread 10 nanometers high with salami filling 5 nanometers thick. They threw mayonnaise—in the form of evaporated metals that stuck only to salami, not to bread—at the side of the sandwich. Viewing the sandwich from the side, they saw defined sideways wires as wide as salami and as long as the edge of a sandwich. They turned their sandwich sideways, transferred the mayonnaise onto a silicon substrate, and went through their mayonnaise spread-and-transfer procedure again to produce the crisscrossed nanowires of the figure on page 154.

An array of metal nanometer wires—scale bar is 500 nanometers—created by researchers at the California NanoSystems Institute.

Courtesy of *Science*)

The researchers had shown that arrays of controllably crisscrossed wires can be created and configured on the nanoscale, letting Nature do much of the work—a step towards a cost-effective means of programming and reading out individual molecular transistors.

Eagerly Awaiting the Marriage of Beauty and Brains

So far molecular electronics has seen both spectacular and debatable results. Diodes and transistors have been demonstrated based on molecules, nanotubes, even single atoms. At extremely low temperatures these manifest the quantum nature of the electrons and show intriguingly complex behavior. But at room temperature, the devices' behavior looks broadly similar to that of a silicon transistor, except that currently, the performance of most molecular devices is dramatically inferior.

So far molecular electronics has seen both spectacular and debatable results.

Molecular electronics needs to show that a sufficient number of devices, seeded by engineers but manufactured by Nature, linked together with a sufficiency of connections, and then programmed, can result in a functioning circuit. While molecular computers will not be designed based on the assumption of perfect, identical chips, researchers must achieve a modicum of control over molecular devices and their connection. The field of directed self-assembly, as it matures, will provide advances crucial to forming robust molecular electronic circuits.

Computing using molecules grown from the ground up, instead of transistors carved from the top down, could result in Moore's Law progressing unabated in spite of seeming obstacles. Much more intriguingly, the ground-up approach could change how computers act and what role they play in our lives. Using molecules efficiently, a 2-pound laptop could perform 10^{51} calculations each second on 10^{31} pieces of information, rather than today's 10^{10} operations each second on 10^{10} pieces of information. This translates into improvement by a factor of trillions of trillions over what we can do today, not just a doubling every eighteen months.

Molecular architecture forces designers into out-of-the-box thinking about the computers of the future. The bottom-up approach mandates new computational strategies in which machines learn and self-correct. Built out of imperfectly arranged molecular neurons, these new computers, like Nature's evolved species, produce masterly results using the simplest—and smallest—of building blocks. Then we might imagine computers built not only by Nature, but by Nurture, their experiences with the world around them reinforcing certain molecular linkages, suppressing others. The resulting machines may have differentiated personalities, built-in potential elaborated by education and enculturation:

eccentric computing. Compact, flexible, organic, wearable: Why not take your molecular computer to the ballet? It might grow up to be the next Baryshnikov. Or it might dedicate its life to the education of the next generation of molecular computers.

8

Interact

Today we interact with information appliances—computers, phones, personal digital assistants—on their terms, not our own. We conform to their requirements mixed in with the legacy of outdated past technologies: the shape and qwertyness of a keyboard, the flat hardness of a display. Why do information appliances not bend to our needs? A tailored computer in a 42 Tall. Today this sounds like science fiction because the materials of electronics are physically rigid, carved from perfect single crystals of silicon. Is it inevitable that information technology and physical cumbersomeness go hand in hand? Will we forever lug heavy laptops, our shoulders de-symmetrized by their weight and awkwardness? Will our laptops always be as expansive as our laps?

No, it does not have to be this way. Powerful computer chips are now so small that their physical inflexibility should play little part in how we experience them. The intelligent part of the computer is a thousand times smaller than a Gucci buckle. Instead it is the screen and keyboard that account for much of computers' weight: the parts that we interact with are made big like us, hand-size keyboards, screens as big as a field of view. But if the parts of the computer we interact with are as big as us, why cannot we make them part of us? Couldn't we tie them to our movements, connect them with our senses? Perhaps we could embed a keyboard in our fingers or weave computers into fabric. Transistors made in polymers could be smooth and flexible, an intergarment between underwear and shirt: true middleware. Download a new thongtone over the wireless Web.

Once we begin molding information technology to our bodies, soon we will want to link directly with our senses. Today 99% of the light coming from a computer screen is thrown away—or seen, for example, by the airplane passenger sitting next to us. Why not make a display that feeds light directly into the eyes—perhaps through a contact lens or a retinal implant? Or cut out the middleman and instead send signals straight to the optic nerve. Then I'll send the signals from my optic nerve over to yours, and finally we shall see eye to eye.

Information appliances today augment our finite human abilities, turning our seven-digit memories into 40 billion bytes, extending the reach of our ventriloquism from 20 yards to 10,000 miles. But there is so much more to do: record every scene we witness for total recall playback; enhance vision into the infrared, in which today we are blind; restore vision to those who have lost it, and give it to those who never knew it. We needn't limit the role of our information appliances to sensing the world around us, but expand their role to involve taking action: a vest that not only measures heart rate, but forcefully contracts to bring a heart-attack

A muscle-suit controlled through your movements and thoughts might amplify your strength a thousand-fold.

sufferer back to life. A muscle-suit controlled through your movements and thoughts might amplify your strength a thousand-fold.

Wearing Your Mind on Your Sleeve

We need materials that can compute, sense, display, communicate, act, and power not based solely on rigid silicon chips, but using flexible lightweight materials that can be integrated into our lives—seamlessly, as it were.

Nanotechnology has already shaped the more conventional ways in which we view information and images. The liquid crystal display in our laptop computers is an example of elegant molecular engineering at work. Liquid crystals are collections of molecules that line up in a defined orientation because of the molecules' shape and electronic structure. How these molecules point relative to one another influences whether light can pass through a series of carefully chosen filters known as polarizers. Applying an electrical voltage to a liquid crystal display pixel changes how, or even whether, the engineered molecules line up, thus determining whether light is transmitted from inside the screen to the viewer's eye. Liquid crystals are a molecular chorus line with perfectly choreographed synchronicity among the dancers.

Advances in carbon nanotubes are enabling a new generation of bright, flat-panel, wide-viewing-angle displays. Field-emission displays operate on much the same principle as conventional televisions—cathode ray tubes—but they require much less power and can be made an eighth of an inch thick. An electric field accelerates electrons towards phosphors, materials that will then glow brightly with a chosen color. Each pixel in these displays is made from many thousands of nanometer-sized tips from which electrons can readily escape. For a given electrical voltage, ultra-sharp tips result in the highest possible electric forces pulling electrons from the tips. Carbon nanotubes provide a natural means of making the sharpest tips imaginable, all through a bottom-up process of self-assembly rather than requiring manual tip sharpening. Samsung and Motorola have both produced advanced prototypes, such as Samsung's full-color 38-inch field-emission displays, based on carbon nanotubes. The displays offer wide viewing angles, low power consumption, and the potential for low cost.

Now the field of organic electronics—the use of semiconductors based on organic molecules, the broad class of molecules from which we ourselves are made—is offering the potential to produce flexible, even wearable, displays. Organic semiconductors can be painted onto a chip, a fabric, or a big plastic sheet. Molecules can be dispersed in a liquid, one that evaporates quickly, the thickness of the thin, smooth film it leaves behind completely controlled. A droplet is placed on a spinning sample, and the droplet spreads out into a thin smooth film. Layers as thin as molecules or as thick as micrometers can be made to cover surfaces in this way. They can also be written into a pattern using ink-jet printing of custom screens, one by one, creating displays tailored to the dimensions of the wearer. Large plastic electronic devices such as the flexible solar cell shown in Chapter 4, Energize, contrast starkly with the rigid crystals of silicon used to make computer chips.

The shape of fluffy electron clouds orbiting molecules determines organic semiconductors' electrical and optical behavior. Electrons must not be tightly bound to individual links along polymer chains, but must instead stretch themselves out in space. Electron waves can be everywhere at once, not just discrete classical particles pinned down to a single location. Just as the long-armed orangutan can travel much more quickly along overhead bars than the Madagascar mouse lemur, electron waves that are omnipresent result in the best conductive materials, for extended electrons move fast and flow abundant current at low energy cost. Plastic semiconductors use long chains of repeating molecular building blocks specially designed to stretch electrons out along their length. Electrons' abundance, in addition to their shape, must also be controlled, and researchers in the 1970s discovered how to put extra electrons into plastics to increase conduction a billion-fold. The demonstration that plastics could be made controllably conductive won Alan J. Heeger, Alan G. MacDiarmid, and Hideki Shirakawa the Nobel Prize in Chemistry 2000.

Making Flexible Displays

Plastic printing technologies have recently been used to make cheap, flexible, lightweight light-emitting color displays—an ideal showcase for what organic electronics can do, and an application not diminished by their existing limitations. Electrons have so far only been made to flow in organics at speeds much slower than in silicon. They are far from being able to compute at the billions of digital calculations per second of which computer chips are capable. But to communicate with humans, none of this is necessary. Computer displays and video screens refresh their images perhaps seventy-five times per second. Existing organic semiconductors can readily operate at these modest speeds.

Today's laptops work by filtering particular colors from a backlight, letting red, green, or blue pass through individually controlled pixels. The images depend strongly on the angle at which these displays are viewed. Plastic displays instead produce light directly within each pixel, and purely of the color desired. And a further approach is possible: produce no light at all, but simply change the reflection of light the way ink does on paper, and thereby alter how light from the room is absorbed or reflected. E-ink, a company that makes displays based on this principle, uses the controlled movement of tiny beads to change the appearance of pixels from white to black. Already department stores are using these as programmable displays.

Plastics that can carry electrical energy are at the heart of a new generation of lightweight displays. In 1989, Richard Friend of Cambridge University found that polymers, when energized with electricity, could be made to produce light. At the heart of the device was a semiconducting polymer layer whose molecules have since been tailored to produce light of a variety of chosen colors with high efficiency. Since different colors correspond to photons with different energies, this meant making polymer molecules that would support different energies of electrons on their chains, just as guitar wire of a different tension will support different acoustic frequencies.

Each light-producing pixel has at its heart a central layer, one that needs to be powered and from which light must escape. The bottom layer, the substrate, is the physical platform on which the device will sit. In a flexible display this is a bendable plastic, one that must be transparent so that the light produced will emerge from the device. On top of the substrate is a sandwich of layers, the bottom and top slices of bread delivering electrical current to the active layer contained between them. The action happens in the light-emitting polymer. A current of electrons is injected, travels under the force created by the circuit, and releases its energy in the form of photons.

Each light-emitting molecule can repeat this procedure millions of times each second.

The light-emitting polymer itself is a triumph of molecular design. Its chemistry determines color of light emission in a way that scientists understand, enabling them to design molecules to specification. Printable light-emitting polymers have been made to cover the range from 400 to 800 nanometers, corresponding to the set of colors our eyes can see. Larger repeat units tend to emit at longer wavelengths of light, more towards the red colors. Another way to tune color unites long-chain polymers with smaller organic molecules that have not linear geometries, but instead snowflake shapes. In the core of these molecules, light is produced. Branches extending from the core control how energy flows into and out of the heart of the molecule; tips on the branches make the molecules soluble, hence paintable on chips, displays—even cell phones, as Philips has recently shown.

Using a similar principle, researchers have wrapped nanoparticles of semiconductor in protective semiconductor shells, creating quantum dots that produce more pure colors than do polymers. Vladimir Bulovic at MIT built a device with an active layer made up of only a single layer of quantum dots, but clad by organic semiconductors: a hybrid device that uses the best properties of each ingredient, his quantum dots producing pure colors, his organics supplying efficient flow of electrons.

The light-emitting polymer itself is a triumph of molecular design.

Once the desired color is produced, electrical energy must be converted efficiently into visible light. External quantum efficiency measures the number of photons emanating from the device for every electron fed in. Important is the underlying efficiency of the material itself in converting electricity into light; the effectiveness of the device in getting electrical excitations into the light-producing layer; and the ease with which light produced inside the device can escape. Choosing strategically the layers of material helps to get electrons into the active layer and keep them there. By optimizing all these features, researchers have achieved external quantum efficiencies sufficient to enable applications in displays and lighting.

Also key is the lifetime of devices. So far, organic electronics has needed to be protected to keep moisture and oxygen out. Devices need to be sealed to perform reliably and efficiently. Even with this, though, the stability of some light-emitting materials remains an issue and, for many applications, needs to be improved. The best polymer light emitters can last for 40,000 hours, well beyond what is required for usefulness in many consumer applications. Not all colors are yet up to this longevity, though: the lifetime of blue-emitting materials continues to require improvement.

Plastic displays are now a commercial reality, having first found their way into devices where light weight and low cost are important. Kodak produces a digital camera with a 5-centimeter plastic electronics view screen. Philips sells a shaver that shows battery charge using an organic display. The organic and polymer molecules serve as pixels in the display, and the electronics that power them are also made from plastic. Concepts are also emerging that take advantage of features distinguishing plastic displays from alternative existing technologies such as the liquid crystal displays used in our laptops. Philips produces a "Magic Mirror" now being embedded in mobile phones. The outside of the phone is a mirror for personal grooming; incoming calls are displayed through the mirror.

The Artificial Retina

Wearable, portable, flexible displays are one exciting new paradigm in human-computer interfaces. We could go farther still: why not cut out the middleman and go straight from the idea of an image to our perception of it? We could use the abundant capacity of our optic nerve to carry information to the mind, bypassing photoreceptors and interfacing intimately with our own electrical signaling.

Our sight is already so polychromatic, sensitive, and highly resolved that it would be difficult to try to compete with our native ability on a first attempt. Instead, researchers at the University of Southern California (USC) in Los Angeles carried out the first such experiments on blind patients whose photoreceptors no longer work. The rods and cones that, in healthy eyes, turn light into electrical signals for subsequent communication with our optic nerves, perceptual regions, and minds, have in these patients lost their function. Retinal degradation, or *retinitis pigmentosa,* affects millions of people: 1 in 4,000 across the population. Many are legally blind by the time they reach their forties. Even more people suffer from age-related macular degeneration, the leading cause of blindness in the Western world.

We could use the abundant capacity of our optic nerve to carry information to the mind.

In these diseases, the optical sensors themselves—the converters of light energy into electrical impulses—stop working, and these patients are left blind. The paths to transport signals from the retina to the brain, though, are still alive: 70 to 90% of nerve structures whose purpose is to receive inputs remain intact and functioning. Researchers have stimulated these neurons in blind patients, producing the perception of bright flashes, showing that the optic nerve has remained active in these blind patients even when optical-to-electrical converters no longer function.

Mark Humayun and colleagues at USC therefore set about building a retinal prosthesis, a system that would include a camera to acquire images and a system to stimulate, using electrical pulses, the neurons of the retina. The team began with building a direct electrical interface to the optic nerve. Their chip was an array of sixteen electrodes and was implanted surgically, placed up against the retina. Further electronics were positioned outside the eye to generate the pulses that would stimulate the electrode array. A separate camera was used to take images and communicate them, via a wireless link, to the electronics driving the implanted unit. The system is depicted in the figure below.

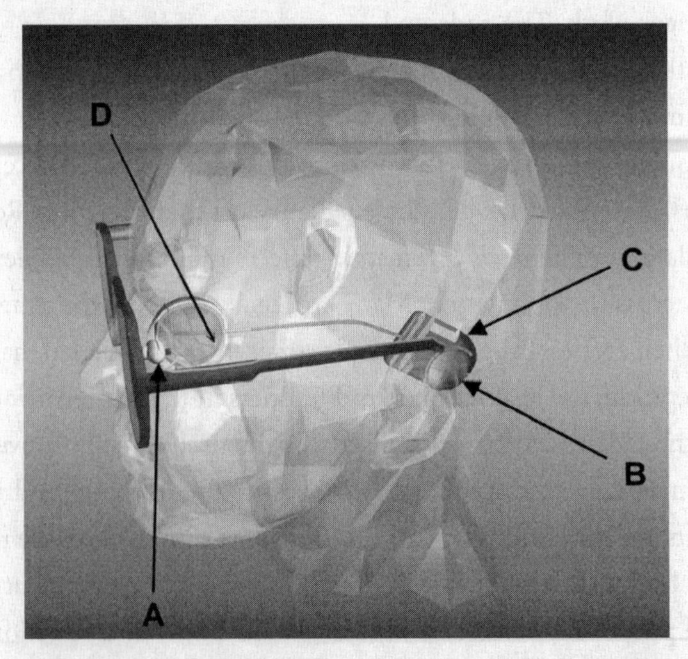

Mark Humayun's artificial retina system. A) Camera embedded in the glasses frame; B) wireless transmitter; C) receiver of signals from the camera; D) electrode array implanted on the surface of the retina.

The external unit consisted of a small camera, worn in the glasses, which captured the scene in front of the wearer. It connected to a belt-worn computer that processed signals and turned measured visual cues into electrical stimuli to be sent to the retinal implant. Signals were fed to the electrode array used to stimulate the retina via a series of wires that crossed the wall of the eye. The system was networked using wireless links.

The researchers' first reports focused on the progress of one patient, a man who had come to them with no light perception in his right eye and only slight light perception in his left. A sufferer of *retinitis pigmentosa,* he had been blind for fifty years. The team first found the level of current in each of the sixteen electrodes that caused the man to experience an equivalent visual perception. These threshold levels varied for the different electrodes, presumably because each contacted retinal neurons with different efficiencies. The group searched for stimulus profiles—the shape of the current over time—that led to the subject experiencing a strong visual response. They always kept the current below a previously determined long-term safety limit.

They then worked with the patient to study his responses to different stimuli. Would he see anything at all? Would he see in color? Might he even form an image? Or might the electrodes unintentionally stimulate other neurons beneath the points of contact, leading to mistaken spatial perception? When the researchers pulsed current through an electrode, the patient perceived a round spot of light, with colors of yellow or white, occasionally red-orange. Sometimes he would see blue when the stimulus was turned off. The man perceived the spots to be the size of a 25-cent piece viewed at arm's length, and his perception of spot location did correlate with the placement of the stimulating electrode. He distinguished stimuli coming from the different electrodes, including from adjacent electrodes less than 1 millimeter apart. He could discern the presence of ambient light and the motion of objects. The subject, blind for much of his life, could now recognize the letter *E* from five feet away. With training and practice, he was able to use the camera to aid his vision.

Humayun is currently trying to enable a blind person to read and perform basic household tasks aided by the retinal implant. Humayun will produce an array with more electrodes crowded closely together, for his first experiments indicated that they had not yet begun to approach the limitations of resolution in their electrode-to-neuron contacts. Their program targets one thousand electrodes compared with our own one million receptors: not enough to drive a car, but sufficient, if effective, to recognize faces.

Integrating Human–Machine–Human Interaction

Nanotechnologists are enriching the links between human beings and the world of information technology. Their discoveries have the potential to augment the interactions we have with our environment and with one another. Flexible displays, ultrasensitive electronic noses, and artificial retinas are but three examples of this progress. Another is Takao Someya's electronic skin, a flexible plastic sheet he and his team at University of Tokyo made that gives a robot a sense of touch. Pressure applied to the robot's artificial glove influences the electric current flowing across the skin, and this current is used to represent the sensation of touch.

People, however, are more than highly sensitive detectors of light, smell, sound, and touch. We are also sophisticated and intuitive perceptors of information integrated across the senses. We rely on a highly interconnected network and a dynamic central processing unit: the brain, which unifies our perceptions. Researchers are working to understand more deeply how we unite the senses—for example, how the synergy between seeing and hearing increases our comprehension well beyond what either cue would allow on its own.

In the meantime, the components that will make up these integrated systems need to be networked. The same idea of ad hoc networks of motes described in the chapter Protect are being used to integrate what is sensed by a constantly changing array of thousands or millions of devices. In future, body-area networks will harvest billions of bits of data each second. Early indications of blood clots forming and releasing, clots that could lead to stroke, will trigger release of blood-thinning agents. I will connect my auditory or olfactory information channel to yours and point to a sound or smell that deserves your attention. The latest news will skim across an unused portion of my field of view, allowing me to read the news with no paper—not even e-paper—just a direct signal to my brain describing the image it would have formed if we still lived in such a concretely physical world.

9

Convey

Today, every long-distance phone call you place travels over fiber-optics, hair-thin strands of glass along which light races at a billion miles an hour. This glass is so free of light-absorbing impurities—the extra atoms added to color stained glass—that light's power is preserved as it flows through micrometer-narrow, thousand-mile-long tunnels. Information-bearing signals keep their shape and sequence even after the long journey: 10 billion playing cards thrown in the air land in order, unshuffled. In communications systems used today, each strand of glass fiber carries one million phone calls at a time. Meanwhile, in the laboratory, fiber-optic systems have been made that carry 100 million calls. Only the inner 20 micrometers of the system needs to be parts-per-billion pure and perfectly cylindrical to the nanometer. This core conveys light without disruption to its trajectory: no friction, no roughness between the fiber core

In communications systems used today, each strand of glass fiber carries one million phone calls at a time.

and its surroundings can be tolerated lest light skid off its intended path.

The optical fiber is a social animal: individual fibers are never found alone in systems. Digging the trench in which to bury it, obtaining right-of-way along railway tracks, and installing the fiber safely in the ground represent a major portion of the cost of putting in a system. And to install hundreds of fibers costs, in percentage terms, little more than to install one. As a result we no longer measure the capacity of fiber-optics in the number of calls a single fiber can carry; this has lost meaning now that one fiber can carry all of North America's traffic, and a modest bundle can carry the world's. Long-haul fiber-communications has been defeated temporarily by its own success, and thus struggling since 2000. Fiber competes only with itself in long-distance communication, and too successfully: 3-cent intercontinental phone calls are the proof.

The need for Internet communications capacity is not limited in the same way as is the need for phone calls. We can only participate in one phone call at a time, so the world's need for voice communication is pretty much fixed. There might be an upper limit to how much capacity the Internet will ultimately demand—something we will not know for a century—but we can be certain that this is thousands of times greater than what phone communication requires. It's harder to characterize a typical Internet session than a phone call—it's like trying to characterize a typical person: one tends to be about 5 foot 8 and 140 pounds, but can vary from 1 to 1,000 pounds, from 1 to 9 feet tall. People are often nice, but sometimes mean, sometimes kind but not friendly, other times friendly but not kind. Internet sessions involve a volley of little e-mails; the BB-gun fire of short text messages; the spritz of an Internet phone call; the spray of a low-quality videoconference; the machine-gun fire of a high-quality videoconference; the random mortar fire of occasional file transfer. And in the future we might like our Internet to carry a fully interactive virtual-reality experience: three-dimensional holographic video, multipoint sound, the re-creation of a sensory experience not limited to the visual and the aural but including the olfactory and the tactile as well. Then a typical Internet session will last between a nanosecond and forever, will transport bits ranging from dozens to infinity, will do so at rates varying from a few to trillions of bits per second, and will demand a level of security somewhere between zero and that required by the CIA.

Communicating Using Light

Researchers in England and the United States strove in the 1970s to perfect the production of pure glass in order to create a pipe for light that would flow optical power over the longest distance possible. They began by finding new ways to remove impurities—atoms other than the silicon and oxygen fundamental to pure glass—below the level of parts per billion. Researchers found how to draw this glass, at elevated temperatures, into long continuous fibers. These needed to be as smooth as possible to form a well-paved avenue to guide light's travels. Smoothness is relative, judged by comparison with what must traverse its surface. A road must be smooth relative to the size of a wheel: speed bumps are felt, but gravel is not. Similarly, since the waves of light are a micrometer in wavelength, glass fibers need to be nanometer-smooth. By the 1980s, long light-guiding fibers carried optical signals hundreds of miles with acceptably low loss of power. This glass was so pure that, rather than being almost perfectly transmissive when an inch thick, as in an airplane window, it was almost perfectly transmissive when more than half a mile thick.

The purity of color in the light traveling down optical fibers is also of critical importance. Pure light is the opposite of white light, which is the mixture of all shades. The deepest tones of blue, green, or red can be pure: like the sound produced by a well-tuned violin, they are a single frequency wave oscillating in time and space. Light travels at different speeds in glass fibers depending on its color. This phenomenon, known as dispersion, makes prisms work and causes rainbows to spread the sun's colors in different directions. In a fiber communications system, if we were to send information on a channel composed of mixtures of blues and greens and reds, the signal would spread out by the time it arrived. What began as a sharply defined parcel of information would turn into a liquid mass of undistinguishable bits, the way previously cold hard pats of butter melted in the sauté pan lose their one-time separate identity.

The use of lasers, generators of the purest light imaginable, minimizes the effect of dispersion. Invented in the early 1960s, lasers were first made from rubies energized using light from a powerful flash lamp to produce pure, deep red. Subsequent inventions in Russia and America led to lasers made of pure, perfect semiconductors, and powered by direct injection of electrical current. The laser relies on two phenomena: gain and feedback. Light's intensity must be amplified as it flows within the laser (gain), and it must be reflected back on

Lasers were first made from rubies energized using light from a powerful flash lamp to produce pure, deep red.

itself as it reaches the ends of the amplifying medium (feedback). Thus high intensities of pure light can be induced to build up inside the laser cavity, and then released controllably.

We've all experienced the conspiracy of gain and feedback in the high-pitched shriek produced when a person using a microphone stands in front of his amplified sound system. Only a bit of sound is needed to get the gain–feedback cycle under way; it is detected by the microphone and amplified (gain) going into the speakers. The signal, now louder, is re-detected by the microphone (feedback), and the cycle repeats itself, growing in intensity. Were it not for the limits on the power the speaker can produce, the cycle would continue until our skin was torn off by the waves of sound; in actual fact, the finitely powered audio speaker can blast only so loud, and we are irritated but our ears ultimately spared. The sound produced tends to be pure of tone even if the murmur that seeded it was a chromatic jumble. How does the speaker know the pitch at which it should shriek? Whichever frequency experiences the most gain and feedback resonates first. The same preferential selection of one most-favored tone accounts for lasers' spectral purity. Laser designers go to special effort to ensure that a single well-defined color is strongly amplified and fed back, further increasing the absolute purity of the light produced.

Today's lasers create light that is pure blue, or pure red, or pure infrared. Their beams are so concentrated that they carve up matter, blast micrometer divots in cornea, steel, and skin. Infrared beams propagating through the atmosphere penetrate fog, sending information through miles of what to our eyes is obscurity. Lasers help measure, to within better than an inch, the distance separating satellites that are thousands of miles apart.

From the point of view of communication, these pure, continuous tones are a beginning, but on their own they contain no information. Light unmodulated is an empty shipping container: information lies in variability. We need to change something about light—its intensity, colors, time of arrival—to convey information, to answer a question. Fiber-optic systems therefore modulate light's brightness, turn a beacon on and off to convey one or zero, yes or no. And as with a beacon, turning the light off all the way comes at a cost: it takes time to start up again, and the filament cools and changes color. A shift in color would change light's speed, resulting in the arrival of a jumble of bits, like sending the first page of a letter by mail and the second by fax. Modulating lasers such that they are not quite extinguished, just dimmed, is a better approach for ensuring their beams carry information clearly.

We can further minimize change in the color of light from a modulated laser by separating the production of light from the imprinting of information. Fiber-optics researchers have created optical modulators, distinct from lasers, whose sole function is to transmit or block light selectably, imparting bits on the channel. These modulators today can put a new bit of information onto light once every 25-trillionths of a second.

These modulators today can put a new bit of information onto light once every 25-trillionths of a second.

As recently as the early 1990s, the light in fibers had to be tidied up every few hundred miles through a rather painstaking process. The signals had lost power, but their shape remained largely intact if the effects of dispersion had been well managed. All that was needed was to amplify the signal as in a stereo system: boost its power while preserving its pattern. At the time, though, it was easier to measure the optical signal, bundle it briefly into a clump of electrons—best guesses as to whether a *0* or *1* had been sent—and then turn it back into light. Instead of using a photocopier to blow up an image, redraw it from scratch. And then came optical amplifiers. Made directly out of optical fiber, they changed how fiber-optics worked. These amplifiers used the fact that energized atoms of erbium—a rarely used character in the periodic table, a high-scorer in atomic Scrabble—boost light. Fiber-optics had settled on a particular infrared color that lost little power and experienced minimal spreading due to dispersion, and erbium atoms impregnated in glass fiber amplified precisely this color, enabling light to gain power along its path, refueling in flight.

This elegant solution reduced further the cost of sending information over optical-fiber networks. Light's full spectrum of colors, though, was still largely untapped at this point. The frequency at which lightwaves oscillate is high: hundreds of trillions of vibrations each second compared with the few hundred from your guitar, the modest millions on your FM radio, the piddling billions from your cell phone. Because lightwaves oscillate at a high frequency, a wide spectrum of channels exists from which to choose, and each one can carry much information. Hundreds of these channels can be accommodated, each one providing minimal loss and dispersion.

Combining many optical channels on one strand of fiber is a technique known as wavelength-division multiplexing and it merged the capacity of forty or eighty separate optical fibers on to one. Before optical amplifiers were discovered, engineers built systems which interrupted the flow of light along fiber every few hundred miles: these would split each fiber's contents into separate channels, detect and retransmit each one, and then recombine the separated colors onto a single fiber. Optical amplifiers changed all of this, allowing eighty or so information-bearing channels to be amplified simultaneously within a single fiber. Practical, cost-saving wavelength-division multiplexing was the result. The merger of multiple new technologies—engineered optical fibers, high-purity lasers, fast modulators, wavelength combiners and separators, and the all-optical amplifier, all optimized to form a unified system—propelled trillions of bits through each fiber every second.

Determining the Limits of Communication Using Light

In 2000, researchers at Bell Laboratories in New Jersey determined a fundamental limit on the amount of information that optical fibers can carry. The work relied on information theory, the mathematics describing the ultimate efficiency of communication. Efficiency is measured in the number of bits of information carried in each cycle: an efficiency of one is excellent, with one bit of information being imprinted on the wave over each period of its oscillation. Depending on how clear the channel is, one could do worse or better than this. The channel is described as noisy if fluctuations mask the correct interpretation of what was meant to be sent. As in conversation, if our surroundings are serene rather than noisy we can understand a great deal more, even making out what is said by the fast, quiet talker.

Communications theorists have developed a variety of techniques to allow a maximum of information to be sent across noisy channels, which are a reality. Sending not just 0's and 1's, but instead selecting symbols from among 0's and 1's and 2's and 3's during each symbol period, can in principle exploit the capacity of the channel beyond one bit per second per wave cycle. Sending with redundancy—in the simplest approach, saying the same thing three times and taking the majority vote—can reduce errors in a noisy environment.

Partha Mitra and Jason Stark of Bell Laboratories explored the information theoretic limits of optical fiber. They recognized that the first obstacle to limit its capacity would be an effect known as cross-phase modulation. To pack fibers as full as possible with information, optical systems engineers would want to cram as many parallel information-containing channels as they could onto one fiber. As long as light travels independently over each channel, and each channel occupies a separate part of the optical spectrum, the signals carried along each can be separated and their information recovered at the other end. Different colors' independence their lack of influence over one another—is a fundamental property of light: two photons crossing paths in outer space—complete emptiness, devoid of all matter—would not disrupt each other's trajectories, and not even know the other was there.

Optical fibers are of course different from outer space—they are tightly packed with silicon and oxygen atoms. Nevertheless, when light is not too bright, photons travel essentially independently of one another down an optical fiber. At higher optical powers, however, one lightwave pushes and pulls on the electrons embedded in glass, and a second wave traveling alongside it feels the influence of these oscillations. Light in one channel thus affects other photons. Keep adding more channels and power and the effect grows: eventually, each channel leaves increasingly unsubtle fingerprints of its own pattern of bits on the contents of the other channel.

Mitra and Stark showed in the laboratory that, even in the presence of this troubling intensity-dependent effect, three bits of information could be sent during each cycle of light's wave oscillation. For comparison, in the field engineers have so far managed to send closer to half of one bit per cycle. Thus we have not yet pushed up against fundamental limits—we're about a factor of six away. But even Mitra and Stark's result had a built-in pragmatic pessimism. The interaction among channels, their mutual imprinting, is not as bad as completely random noise. Since we know how photons interact with one another as they propagate inside optical fiber, we can in principle figure out, from the jumble received, what must have been sent. A smoking gun, a fingerprint, and a body may not prove incontrovertibly that a crime took place, but coupled with a propensity and a narrative, they may suggest it beyond reasonable doubt. Communications engineers regularly infer what was sent based on the distorted signal received using their intimate knowledge of the communication channel over which the signal traveled.

Each one of us routinely does the same sort of work as these researchers. Our view of the *Mona Lisa* at the Louvre is obscured by reflections from the eleven panes of bulletproof glass bouncing light from forbidden camera flashes. We see the juxtaposition of the *Mona Lisa* and flashbulbs and pink HELLO KITTY: JE SUIS LÉGÈRE COMME TROIS POMMES T-shirts. In our minds, in an instant, we subtract the distortions and see the work of art, in part because we have defined expectations of what we are to see. Instantly we solve a complex inversion problem.

Mitra and Stark's result is a useful guide. Should further techniques be deployed, such as multilevel coding and phase modulation, the information-carrying capacity of today's fiber-optic networks could be further increased by a factor of about ten, saturating once we approach about 12 trillion bits per second. The end of leaps-and-bounds growth in optical fiber's capacity is within sight.

Networking with Light

Advances in optical communications systems have been measured using a single quantity: information rate times distance. The value of the Internet is much harder to pin on a single quantity—it's certainly much more than one fixed, ultra-capacious connection. The true value is in what we can do through its ever-changing links, creatively using alternative paths for information not foreseeable when the network was built.

The value of the Internet is in what we can do through the ever-changing use of its links.

The U.S. hub-and-spoke airline routing system is another example of a network. It's infeasible to have direct flights, fixed point-to-point links, from everywhere to anywhere. To connect each of 100 airports—Rochester, Minnesota; Northwest Arkansas; Jackson Hole; Boise Intergalactic—with 99 other cities would mean nearly 10,000 distinct routes. The airlines are in enough trouble as it is. Instead, the airlines route through hubs—Chicago, Dallas, La Guardia, LAX—to get passengers from one destination to another. This ensures that the number of separate routes needed scales instead with the number of destinations. One benefit of this network architecture is redundancy: there is more than one way to get from Boston to Orange County—travel via Dallas, Chicago, or Denver, for example. One of these—Denver—is the shortest routing and thus the preferred. But if the Boston–Denver or Denver–Orange

County flight is full, then the alternative routings come in handy. There exists a built-in way—alternative routings—of handling overflow along a given link. This redundancy not only helps with flow, but also makes the network better able to withstand disruptions. If Chicago is taken out by a snowstorm, a number of ways remain to travel between Boston and Seattle.

This networked approach enhances efficiency by reducing the number of flights. It does, however, put a new burden of responsibility on the efficient switching of passengers within hub airports. But what actually goes on in airports is strikingly unwieldy compared to the graceful arc of the modern jetliner. Passengers undergo a sudden transformation: in the sky they were soaring at over 550 miles per hour, just shy of the speed of sound, but in airports they are turned into slow-walking, giant-pretzel-eating pedestrians.

The situation is even more absurd in fiber-optic communications systems. Bits traveling down optical fibers race along at the speed of light, about one billion miles per hour. But because the airports of the Internet—switching terminals known as routers—today function electronically, soaring photons turn into dawdling electrons an average of fifteen times along their path. These sit on uncomfortable benches known as electronic buffers, waiting for the next supersonic leg to begin. The waiting areas are getting crowded: information networks handle petabits—1,000 trillion pieces of information today—and our enthusiasm for digital travel over the Internet is increasing. Optical networks need a more elegant solution. Dubai International Airport advertisements depict the ultimate fantasy: no long, stationary airport where planes land and passengers sit, but instead a flying docking station. Planes attach and detach, passengers flow through the core of the airport down and up fast-moving escalators into their next plane. No one stops moving at plane-speed to make their connection.

Optical networks researchers are beginning to use nanotechnology to do the same. Information-bearing photons would no longer be decelerated to electron speed, shuffled off to their new departure gate, made to wait for the next flight, and eventually fired off in a new direction. Instead, photons would fly into a node and be instantly deflected into the next fiber path along their trajectory. In these all-optical networks, signals would remain photonic over space and time. The ultimate in optical switching means shuffling light as fast as the bits are coming: tens of billionths of seconds today, perhaps even faster in the future. Nanotechnologists are making the materials and devices that will act in one-trillionth of a second to reconfigure the direction in which light is deflected at the node within a network.

Bouncing Light through the Internet

Making light do more than just carry information along fixed paths, but instead skip nimbly to any destination, requires us to alter photons' directions with exceptional control. We need materials that make light interact strongly with light, and from these we must build devices in which one optical signal controls another. These devices must route signals without causing too much photon attrition—loss of power in the signal to be switched—just as the rate of plane crashes needs to be kept below a certain threshold. Nonlinear optical materials must perform their magic on a specific set of colors, those in the infrared in which light travels unfettered down optical fibers. Optical networks demand designer devices made from materials that exhibit a set of properties not found in Nature: nanotechnology is the ideal candidate to fill the void.

Buckyballs, 2-nanometer carbon soccer balls, have recently been turned into materials for high-performance optical switching. In 2000, theorist Mark Kuzyk at Washington State University used quantum mechanics to compute how strongly light-switching a material could be. His mathematical derivations complete, Kuzyk compared his results with what other researchers had managed to achieve in the lab and noticed a striking gap. Of dozens of reports published in the literature, none came closer than within a factor of thirty of the fundamental limit Kuzyk had derived. It was as if someone, using knowledge of human physiology and aerodynamics, had determined that humans should be able to run a four-minute mile; but nobody had ever managed to run a mile in under two hours. What were we doing wrong? The cost of underperforming by a factor of thirty was severe: signals could not be switched except when they were conveyed using impractically high intensities of light.

Kuzyk used his theory to suggest how a chemist could build a molecule that would achieve its full potential, and a team of experimentalists set about turning Kuzyk's dream into reality. Synthetic organic chemist Wayne Wang at Carleton University in Ottawa, Ontario, would implement the vision, be the molecular dressmaker. Researchers in my group at the University of Toronto supplied the engineering, designing, tailoring, and measuring for fit, looking critically at the new material's performance as judged by practical engineering rules for optical switching. The chemists began with fluffy electron-rich clouds of buckyballs, but recognized that in their pure state these molecules would not do the trick. Guided by Kuzyk's principles, we combined the buckyballs with a polymer, attaching it to the buckyballs' surface, and using this new control to shift buckyballs' resonances out to the colors used in optical communications. Ultimately, through rational molecular design and realization coming together across chemistry, physics, and engineering, we came within a factor of two of the fundamental quantum limit. It was as if we had been forever running four-minute miles, and all of a sudden someone had run a mile in sixteen seconds.

Trapping Light

There is a further dimension critical to making devices that will switch light using light: it is necessary to transform a material's subtle changes in properties into a bold, unambiguous deflection of light from one fiber into another fiber, launching it onto a new optical path towards its intended destination. Microcavities are cages for light that trap photons inside their minuscule volume, inducing huge numbers of photons to pile up in a small container. Like a highly resonant tuning fork that reverberates at one pure pitch alone, microcavities are highly discriminating in their treatment of light depending on its color.

Trapping light is useful, for it allows enhanced control over its behavior. It takes considerable effort, though: light yearns to be free. This distinguishes it from electrons, which, even if accelerated to great speeds, will return rapidly to captivity of their own volition. Light would rather either keep on going at 300 million meters per second, or, if it is to be caged, will often be absorbed, its vital energy lost. Thus the challenge is to keep it alive in captivity. Light is a panda.

Trapping light is useful, for it allows enhanced control over its behavior. It takes considerable effort, though: light yearns to be free.

Researchers have recently trapped light inside the smallest scale imaginable—that of light's own wavelength, a few hundred nanometers. Yoshio Yamamoto of Stanford University in California has built extremely reflective mirrors and has paired them up such that light is trapped between them, as shown in the figure below.

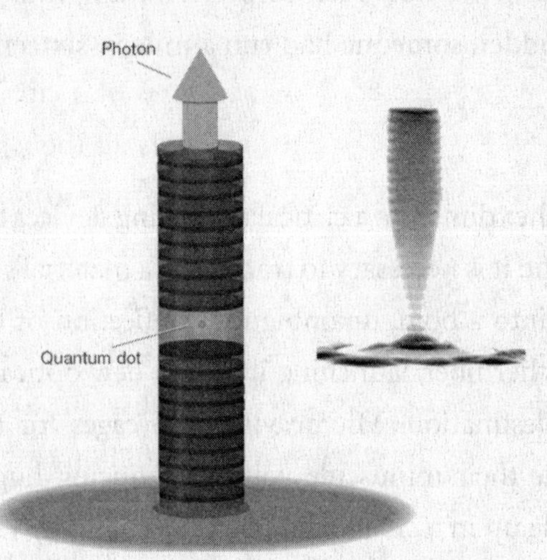

Micropost cavities that confine light to a small volume. (Courtesy of *Nature*)

Light's energy stays trapped within these devices for a long time before it escapes, just as the tuning fork resonates long after it was excited. The trapping occurs by means of ricochet: waves may scurry back and forth at high speed, but their energy ultimately stays put. Microcavities are important not only in switching, but also because of how they transform the behavior of light produced within them. Yamamoto made single-photon sources in this way, machine guns that fire individual photons with perfect regularity. A classical light emitter—a light bulb for example—emits photons randomly. Yamamoto's device produced photons one at a time upon command: a noiseless source of information-carrying photons.

Kerry Vahala and his team at Caltech in Pasadena, California, have built not a cage for light, but a racetrack. Vahala's spherical and circular structures both rely on whispering-gallery modes: instead of bouncing back and forth along an axis as with Yamamoto's mirrors, Vahala's light orbits in defined circles, chasing and catching its own tail. The devices recall the acoustics of a circular theater in which a murmur up in one balcony travels all the way around the gallery, touching each audience member sitting around the periphery, eventually returning to the original speaker as an echo.

The first successful optical demonstrations of this idea used perfect glass spheres. Recently, Vahala reproduced his success in a much more convenient structure, a flat disk on a chip. Photons launched inside were stored for over 100 million wave oscillations. It was critical for the device to have a top surface smooth on the scale of atoms, something Vahala achieved by taking the disk through a molten state during fabrication, enhancing the reflectivity of his surface just as torching crème brûlée smooths rough sugar's scatter into a smooth mirror finish.

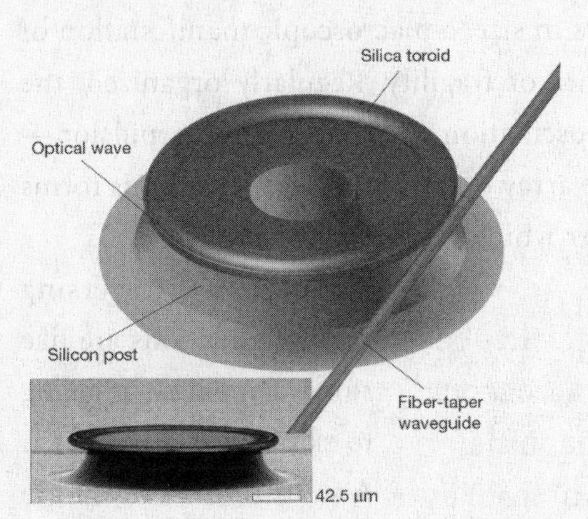

Vahala's microtoroids, carved in glass, and connected to the outside world by gentle coupling to a bared optical fiber. (Courtesy of *Nature*)

Microcavities have been created to enhance dramatically our control over the fate of light. Compared with our dexterity in working with electrons, our ability to trap and manipulate photons has only begun to fulfill its tremendous potential. In integrated circuits we can connect hundreds of millions of transistors, and this complex interconnection of simple functional elements has allowed continued exponential growth in the sophistication of electronic computers. To date, photonic devices have largely been made one by one, then interconnected individually. Could we instead create a semiconductor for light? Researchers are seeking a platform technology for refined, integrated control over photons.

Building a Semiconductor for Light

The properties of semiconductor crystals in controlling the flow of electrical current account for their spectacular success in giving rise to the information revolution. Central to their behavior is the regularity, or periodicity, of their array of atoms, packed like eggs in a carton and then layered vertically as well. Crystals crush into perfect cubes millimeters in size, a macroscopic manifestation of nanometric atomic planes of fragility. Regularly organized, the atoms create a periodic oscillation of attraction and repulsion—electrons see a harmonic array of hills and valleys, and this forms the rolling landscape over which they must travel.

Crystals crush into perfect cubes millimeters in size, a macroscopic manifestation of nanometric atomic planes of fragility.

Electrons traversing these undulations are like stones skipping—or failing to skip—over wavy water. A stone thrown at just the right speed will skim effortlessly from crest to crest. Thrown too fast or too slowly, it will become trapped between waves and soon sink into a trough. In semiconductors, electrons that have just the right energy could skip over the crests of waves nearly forever. Electrons lacking this resonance will quickly become trapped, luckless in their quest to surf from Cuba to Miami. Electrons forbidden to propagate are said to lie in the bandgap—conduction's purgatory.

Launch electrons that readily skim the tops of waves and you have a conductive material. Launch electrons such that they crash head-first into vertical walls of water and you have an insulator. In between you have a semiconductor whose conduction can readily be tailored over a wide range. Semiconductor engineers make some regions conducting, others insulating, and combine those to make transistors. Photonics researchers, envious of their electronics brethren, asked in the 1980s whether they could gain similar control: create a medium within which to exercise influence over the propagation of light. In 1987, the field of photonic bandgaps was born. Two researchers simultaneously published the idea of a photonic bandgap in the preeminent physics journal *Physical Review Letters*. Eli Yablonovitch was then at Bell Laboratories, soon to launch his career at UCLA. Sajeev John was finishing up a post-doctoral fellowship at Harvard, about to take up a faculty position at the University of Toronto.

Electrons forbidden to propagate are said to lie in the bandgap—conduction's purgatory.

Yablonovitch and John both used the fundamental mathematical theory describing light to show that photons, like electrons, could be trapped in a bandgap. They compared the equations that describe how electron and photon waves flow. Both equations began by depicting the undulations of their waves and finished by representing the waves' energy. Sandwiched in between was the external undulation, the shape of water waves over which pebbles would skip. In semiconductor crystals this was the landscape—the hills and valleys—seen by electrons inside the crystal. In the case of photonic bandgaps, the density of the materials from which the metaphorical crystal was sculpted would govern the fate of light.

Yablonovitch explored the analogy in a way semiconductor-chip researchers would understand: when packed into perfect arrays atoms produce bandgaps for electrons, and similarly, perfectly periodic arrays of highly contrasting optical materials should create a bandgap for photons. John looked at the question not through the lens of perfectly regular crystals, but through that of choppy, irregular waters. He showed that if the amplitudes of the waves were great enough, light would get trapped within about one wave's separation.

The theory of photonic bandgaps thus doubly conceived, who would deliver the idea into the world, be the first to prove it experimentally, trap light inside three-dimensionally periodic, high-contrast materials? Researchers divided into two camps, two ideologies: the top-down and the bottom-up. The top-downs would carve shapes from preexisting sculptural blocks, revealing the photonic soul of their soapstone; or they would build perfect egg cartons and pile them on top of one another, all with precision much finer than the wavelength of light to be trapped.

In 1998, researchers at Sandia National Labs in New Mexico built, layer by layer, woodpile structures made not of wood but of silicon. Because of the nature of the lithographic process, each layer had to be separately specified in photoresist; its pattern transferred to the silicon beneath; the layer made planar to allow the next layer in the woodpile to be laid out; and the procedure repeated, with suitable alignment of all subsequent layers relative to the original. The Sandia labs' work provided a photonic bandgap at a wavelength about ten times longer than the colors used in communications.

Andrew Turberfield and his colleagues at Oxford University developed a method to build photonic crystals not layer by layer, but in one single, three-dimensional flash of light. No knife, but rather a hologram, would carve an intricate structure. Turberfield sculpted photoresist, plastic that can be dissolved once exposed to light. He set up a series of laser beams, launched in at different angles, which converged inside a thick photoresist layer. Upon meeting, the beams added up to form a super-wave when crests coincided, or cancelled one another when trough met crest. Turberfield created thereby a complex pattern of light intensity and imprinted it on his polymer. When the soluble bits were washed away, what remained was a photonic crystal.

A recent triumph for the top-down camp came from the efforts of Hank Smith and collaborators at MIT. This structure was built up carved-layer by carved-layer. The structure used free-standing pillars 70 nanometers in size—incredibly narrow features to construct reproducibly—aligned, over many layers, all using electron-beam lithography. As with all successful efforts in the area, the work united researchers who could design new photonic lattices with those who could build and study them. At this level of sophistication, design and fabrication cannot be separated—it is a simple matter to come up with an un-buildable photonic crystal, but a challenging one to invent a lattice the nanobuilder can actually construct.

Among the bottom-ups, Sajeev John, Geff Odin, Henry van Driel, and their collaborators in Toronto and Spain had a major success in 2001. In the years leading up to this achievement, the bottom-ups had been coaxing Nature into organizing innumerable identical glass microspheres into regular, closely packed arrays—increasingly perfect self-organized photonic crystals. These did not, however, exhibit a full photonic bandgap—a range of frequencies in which light, incident at *any* angle, was forbidden—at the wavelengths of light used in fiber-optic communications. Instead, these photonic crystals altered light's experience, but it could always find a way to escape. The lion that can escape by walking out the side of the cage is not a well-caged lion. The researchers needed to build a material whose constituents contrasted more starkly in their optical properties than did glass and air. They would also build a topology that Sajeev John had predicted would be well-suited to trapping light. Researchers from Valencia and Madrid had perfected the means of making the template: a lattice of glass spheres of just the right size and spacing. The group heated their crystal, fusing glass spheres where the beads touched. Odin and colleagues filled the holes in the structure with silicon, a much more optically dense material than the glass spheres. They used a chemical to remove only the Spanish lattice, leaving the inverted crystal made out of silicon. Van Driel's team showed that the resulting structure made the light used in communications dance on command.

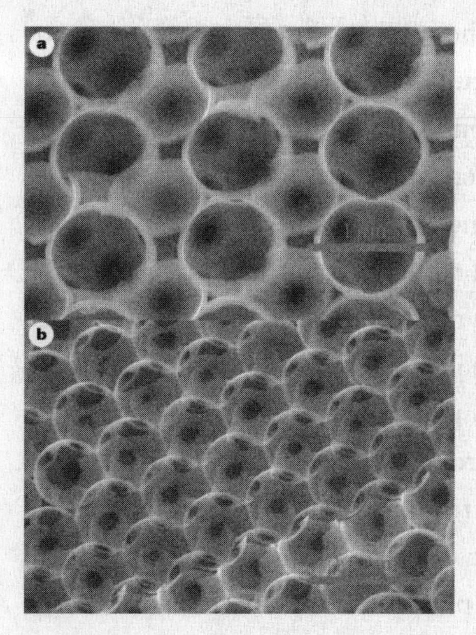

The photonic crystal made by the Toronto–Valencia–Madrid team. Spheres fused where they touched and created a structure that was converted into a photonic bandgap medium. (Courtesy of *Nature*)

Integrating Light with Electronics, Communications with Computing

Networks based on light are not self-sufficient—they are fed with data from computers, cell phones, and many other information appliances. The same could be said of our international air travel network in an earlier analogy. Charles de Gaulle airport cannot exist on its own, but is elegantly integrated: *trains à grande vitesse* run underneath and passengers can go from planes onto Paris's system of bicycle paths.

Since the Internet is everywhere, the same integration and interconnection are needed among networks. Everyone's home is wired; and the wireless network of cell phones, Blackberries, PDAs, and WiFi in laptops makes the Internet ubiquitous. Today, the links are awkward between wireless devices, home and office electronic networks, and fiber-optics for long-distance communications. Our computers hook in through their silicon-integrated circuits; our cell phones use specialized circuits made using gallium arsenide, a more costly and exotic semiconductor incompatible with silicon; and the lasers that run the fiber-optic network sit on yet another type of costly crystal. These home, wireless, and optical networks are cobbled together. Our physical reality has not caught up with the convergence of our networking needs.

The need for convergence makes it important to turn electronic materials such as silicon into a photonic material. Silicon is, in its native state, not an efficient producer of light. Electron waves do not readily come together and release their differences in energy in the form of a photon. But researchers discovered in the 1990s that silicon could be modified to produce light more efficiently. They degraded the perfect single crystal of silicon, creating nanometer-size crystals for electrons to inhabit, and from which electrons spontaneously produced light. In 2000, a group at the University of Trento in Italy observed optical gain—the key ingredient in making a laser—from silicon nanocrystals. Two years later, a group at Intel, using the techniques already available to produce Intel's Pentium processors, showed that they could create an optical modulator—a device that puts information on to light from a laser. The silicon-based modulator imprints the message contained in an electrical signal on an optical carrier. Intel's modulator brought together the best of electronics—its manufacturability and low cost—with the speed and functions required of optics. And in late 2004 and early 2005, first the group of Bahram Jalali of UCLA, soon followed by

Our physical reality has not caught up with the convergence of our networking needs.

the team led by Mario Paniccia at Intel Corporation, made lasers using silicon.

To defy the incompatibilities of different semiconductors is also possible, making forced marriages between crystals whose atoms are differently spaced. This approach has seen considerable recent success. Gene Fitzgerald's group at MIT has shown that the semiconductor gallium arsenide used in cell phones can be built successfully on silicon, in spite of the two semiconductors' different atom-to-atom separations, without compromising performance. Fitzgerald used buffer layers, intervening crystals in which the silicon and gallium arsenide work out their differences. The semiconductors agree to put all of their disagreements into this intervening layer. The approach is promising as long as, over time, the differences can kept inside the no-man's land, rather than leading to border skirmishes.

Recently researchers have put optical materials on electronic ones without matching crystals to one another at all. Instead of growing the optical crystal on top of the electronic one, researchers are making the optoelectronic materials in a liquid, swaddling them in a protective molecular wrapper, and then spreading the resulting paint overtop the silicon. The liquid used to disperse the quantum dot nanoparticles evaporates, and what remains is a film of well-encapsulated nanoparticles. Collectively, the teams led by Uri Banin at Hebrew University of Jerusalem; my group at the University of Toronto; and the collaborators Vladimir Bulovic and Moungi Bawendi at MIT have reported significant progress since 2001 along this path. Each group has made a source of light for communications on-chip or within networks. The Toronto group recently reported that we were able to make a photodetector—a device that measures the infrared optical signals carried over an optical network—with promising levels of efficiency. Teams at Los Alamos National Lab and at the University of Toronto separately reported the observation of optical gain—necessary for lasing—at wavelengths needed in communications, all based on films of nanoparticles paintable onto a silicon chip.

Bringing about the Death of Distance

What if we could unite electronics, photonics, and the translator between the two—optoelectronics—conveniently on one chip? Light could then become used everywhere it is advantageous, not just within optical fiber as today, but to connect chips to one another and even link subunits within a chip. Power consumption and elevated chip temperatures are the most urgent factors limiting progress in electronic integrated circuits, and using light as a communication medium on-chip is a possible solution—able to convey a maximum of information while consuming a minimum of power.

Networking using light, combined with the seamless interconversion among electronic, wireless, and optical communications formats, means more than just the gradual decline in the importance of physical distance that we have seen in recent decades: cheaper long-distance phone calls and choppy video over the Internet. Ultimately it will lead to the death of distance. If two computers physically situated halfway around the world are, from the standpoint of sharing information, as close together as two computers sitting in the same room, then the world's set of computers can be merged into one massive super-computer.

We've had a first taste of such global computational reunification: screensavers that search, while the computer is not otherwise in use, for signals betraying extra-terrestrial life. This is the beginning of grid computing, in which the world's computer resources become one huge pool. Eventually computers will be so inter-networked that no one will lack the capacity to process bits. Instead, we will off-shore our thinking: drug discovery, for example, may be accelerated, with new protein structures conceived and studied for cancer-cell binding all in an instant and inside a computer wrapped around the globe.

We have already seen tempting signs of what may be possible. The University of Pennsylvania developed an electronic medical-record data grid and repository, one that captured, from any location, a patient's full range of healthcare files: high-fidelity medical images, records, and clinical history. Previously, the chain of records was often broken, for example, when patients obtained mammograms at different locations. The new system let doctors see all relevant records within two to ninety seconds and treat their patients with a new level of efficient, responsive care. Just as an electrical grid, which delivers electricity to homes, equalizes our energetic experience independent of geography, computing grids level the knowledge landscape when abundant, valuable information can be anywhere it's needed in a flash.

This is the beginning of grid computing, in which the world's computer resources become one huge pool.

The death of distance could apply not just to computing systems and data repositories, but to interactions among people. The end of business travel has been heralded for decades, but today, even in the world of e-mail, videoconferencing, and phone calls, people still fly across continents to meet in person. Why? In today's world, distance is still alive and kicking: we have not yet used technology to create the sense of presence—a touch, a smell, the feeling of being there—from afar. Even if we could represent sensory experience so persuasively, we wouldn't have enough networking capacity today to send these multiple terabit-per-second experiences anywhere in the world reproducibly and instantaneously. A smart wall will need to depict the Tokyo side of the desk in the virtual conference room with perfect acuity. Holograms will need to express a three-dimensional, lifelike presence, and sensory suits will need to translate touch across thousands of miles. Once achieved this will mean no more excuses for us to show preference to our co-located colleagues. The world's talents will become a unified resource.

Friends should no longer need to live in the same place: cycling buddies living in Boston and Toronto should pedal together with the feeling of closeness reproduced, integrated, and instantly conveyed. Executives should no longer need to fly across continents to play squash and carry on affairs. They will merge relevant real-time descriptions of physical reality without being forced to unite the physical presence of their atoms. Two carbon atoms are the same irrespective of the coast on which they lie, as are two oxygen atoms. All that really matters is the information describing their relative configurations, energies, and surroundings.

This view places information above matter, and it supports a vision quite distinct from rebuilding Greta Garbo from the bottom up. Instead it re-creates experience across space and time, shares imagination among people and places, capturing, conveying, and expressing these representations with finesse.

But must the molecular and informatic views of the world stand opposed to one another? Not at all—nor indeed can they. As with light and electrons in networks, literalism and metaphor in prose, experiment and theory in science, rigor and creativity in all disciplines, true command over our parallel universes—the conceptual and the tangible—will be gained only once we see and harness their complementarity. For the dance of molecules to be anything other than a mosh pit, we need inspired choreography.

Epilogue

Humanize

We have roamed chemistry, physics, biology, and information. We have seen how electrons, photons, atoms, and their assemblies may be applied to solve problems in health, the natural environment, and our interactions with one another and with human-made machines. What linked these disparate themes? They were not unified by a single guiding scientific principle. No puritanical dedication to the nanometer excluded all other ideas of interest. No dogmatism insisted on quantum effects alone. Neither bottom-up chemical self-assembly nor top-down carving of matter won the day.

The diversity of topics visited in *The Dance of Molecules* illustrates that nanotechnology is not a single set of scientific principles or applications. Is it, then, a coherent intellectual enterprise? More culture than content, nanotechnology is a fresh outlook, a reevaluation of what is to be cherished in the process and purpose of science and engineering.

Nanotechnology is an active search for convergence among traditionally disparate disciplines, a quest to see the unity of ideas. Computer network researchers, equipped with motes, seek inspiration from swarming bees. Semiconductor physicists pan for biological proteins that will allow refined control over nanoparticles in early detection of cancer. Ophthalmologists exploit chipmakers' control over silicon to return sight to the blind.

Nanotechnology is a rebirth of Renaissance science in a time when to be a true Renaissance scientist is no longer possible. The tree of knowledge has grown too many branches: no individual researcher can be simultaneously the leading expert in physical chemistry, electronic systems engineering, and cancer biology. Fortunately, though, within this new framework a researcher's discoveries are not limited to what he can fit inside his own mind. If he can find partners with complementary skills and knowledge, together they can assemble the expertise needed to carry on work of exceptional quality and practical importance.

Nanotechnology is a rebirth of Renaissance science in a time when to be a true Renaissance scientist is no longer possible.

Communication of nanotechnology's possibilities, methods, and achievements then becomes more essential still. It has always been critically important: researchers have long owed a debt of gratitude to their patrons, who today are the taxpayers that invest in fundamental and applied research. Without society's investment, researchers could not experiment in the lab all day and night; they could not toss around ideas crazy and sane for reflection, reformulation, and rebuttal. The highly networked nature of nanotechnology's advances augments the importance of broad communication of science's newest ideas. Only if we explain to other researchers, designers, politicians, translators, landscape architects—creative, curious people in fields close and far—what we are doing and why it could be important, will we supply them with the ideas and knowledge they need to work together with us towards a shared purpose.

In the 1990s, Bob Metcalfe, inventor of Ethernet, founder of 3Com, coined the following law: *The usefulness of a network grows not in linear proportion with the number of users, but with the square of this number*. The telephone network was not a network when two friends each had a phone. When each of us came to have a phone in our home, the globe became our neighborhood. Now that we each carry a phone everywhere we go, our global neighborhood has become a fluid cloud of omnipresence. Our voices and ideas and emotions diffuse through the ether at our command. The same thing happens—isolated thought transformed into creative explosion—when separate constellations of new ideas are unified in an effective network. Creative critical mass. When chemical biology is the closed domain of chemical biologists, its innovations and progress are limited by the circulation of ideas within a fixed community. Communicating with a diverse set of physical scientists leads to designer DNA being used to grow custom nanoparticles that light up cancer cells at the wavelengths where our tissue is transparent. If these discoveries waft across the research community, including among medical practitioners, we may learn how to translate these new concepts into clinical reality. Further, such discoveries are more likely to be embraced rather than resisted. If researchers speak out early about their emerging concepts, then philosophers, ethicists, and policy-makers, among others, can anticipate the risks and benefits—intellectual, moral, and societal—of scientific breakthroughs. And they can take action before an avalanche of infinite possibility engenders fear.

Nonagenerian Suffers Tragic Premature Death

In the future we will increasingly detect and, using targeted delivery of drugs, arrest cancer before it spreads. Through regenerative medicine we will replace worn cardiovascular components before they fail us. Life expectancy at birth in North America rose from forty-eight years in 1900 to seventy-seven in 2000. Suppose that we could not only maintain, but accelerate, this pace. Would the average lifespan increase to well over one century in only fifty years? If so, would quality of life be improved commensurate with quantity? The goal of life extension must be not to make us older longer, but younger longer.

The cost of quantity with no enhancement in quality is devastating. Alzheimer's has exploded in recent decades because of its late onset combined with greater life expectancy. Three percent of people ages sixty-four to seventy-four suffer from Alzheimer's and almost half of those older than eight-five years are sufferers. A friend who works in the hospital reports that today's heart patient holds on long enough to experience a much richer cocktail of syndromes than he would have some years back. Where once the patient's heart attack would have killed him, today he survives the heart attack and lives long enough to experience stroke, undergo renal failure, and then suffer ischemic leg pain.

Will we tack on thirty years of life, but not of living, at the end of our time on Earth? Fortunately the same advances that extend life tend to improve its quality. Stem cell research and molecular medicine could treat Alzheimer's. New designer drugs target arthritis, cancer, cardiovascular disease, and stroke. Tissue engineering produces bioengineered bone, blood vessels, muscles, and nerves to repair or replace organs in ageing bodies. And there is a precedent for life-extending technologies to lengthen quality of life as well: the sixty-year-olds of today have the same health as the forty-five-year-olds of 1900.

Suppose then that duration and quality of life can both be extended considerably. What will this mean for society? Life cycles will change further. Our choices of when to study, spend, marry, reproduce, work, and retire will be influenced by how many years we think we have left. Just as you might live your life differently if you discovered you had one month left, so might you change your style if you found that you had another eighty years in you. No more sunny days without SPF 80. Spending habits might change in those intent on feeding themselves to age 140. Recently the age range for childbearing has become compressed: women delay having children but then face with urgency the fact of reduced fecundity as they reach their late thirties. Further advances in ex-utero reproductive technologies may enable women to have children—at acceptable cost and risk—into their fifties, sixties, and seventies. The choice to have children would be expanded across the span of life, the childbearing years thus decompressed. The implications on social dynamic are tremendous. Today, the advantage in courtship lies in the third decade of life with women and in the fourth decade with men. Age-agnostic reproduction could even out the situation.

Our choices of when to study, spend, marry, reproduce, work, and retire will be influenced by how many years we think we have left.

The financial implications are huge as well. A sudden jump in longevity could mean a generation of inheritances lost as great-grandparents survive into their 130s and exhaust their supply of funds. Current pension schemes, which have bet on the average pensioner dying within fifteen years of retirement, would prove to have been hopelessly under-funded. Future generations of earners and taxpayers would not have the option of retiring anywhere near sixty-five years of age. This may suit them, for life at sixty would be the new halfway point, and their ambitions perhaps nowhere close to being fulfilled.

Today, education is viewed as a lifelong project. This will become increasingly true as science and technology advance at a rapid rate. Imagine—in this, the hundredth anniversary of Einstein's discovery of the photoelectric effect—if a student born in 1885 had finished his formal education by 1905 and were still contributing scientifically today at age 120. He would have had to pick up along the way, as they emerged, all of modern physics, semiconductor technology, and computing. We and our children will in future be required to learn much, much more over our lifetimes.

Military Power: Technology Gap Widens

Many of the technologies discussed in this book have military applications. Designer organs super-endow soldiers with disproportionate physical prowess. Active battle suits protect soldiers from nerve gas and resuscitate them when wounded. Portable solar tents silently capture power in the desert. Smart dust sprinkled over enemy territory invisibly reports the minutiae of enemy habitats and habits.

The U.S. military is among the world's leading investors in nanotechnology research. Within the U.S. National Nanotechnology Initiative—the ensemble of federal investments in nanotechnology—

military funding accounts for about 30% of total spending. The United States is the dominant spender in military R&D globally, accounting for two-thirds of all such spending worldwide, $52 billion in 2002. Every service within the U.S. military has programs in nanotechnology. High-profile activities include the U.S. Army's investment of $50 million to form MIT's Institute for Soldier Nanotechnology (ISN), augmented by an infusion of $40 million from industry. The Naval Research Laboratory has formed an Institute for Nanoscience.

Publicly discussed investment in military nanotechnology tends to focus on defensive technologies, especially those aimed at protecting the soldier in the battlefield. The Institute for Soldier Nanotechnology's guiding vision includes a battle suit that protects the soldier against bullets ("energy-absorbing material"); that can, on command, protect the soldier from chemical and biological agents ("switchable nanopores"); that provides integrated communications and networking; that can change color to provide camouflage; that can apply force either to aid the wearer in lifting loads ("muscle-suit") or in compressing wounds; and that can sense the soldier's vital signs and act on this information, such as by delivering drugs when needed.

These technologies aim to protect human lives. They could enable peacekeeping efforts that are more tractable, assisting developed nations to enforce international codes of human rights with lowered risks to their own personnel. At the same time, though, by decreasing the vulnerability of the soldiers of affluent nations, such technologies hold implications of an offensive nature. Any advantage in the battlefield shifts the balance of power. The Western world's point of greatest vulnerability in any military decision is the huge value it places on the lives of its own combatants. The U.S. military budget request for fiscal year 2005 is $420.7 billion, yet national concern focuses not so much on spending as on the cost to American soldiers' lives. The global dynamic could shift if the United States were able to act in any military environment—including interventions on the ground—at drastically reduced risk to its soldiers' lives. Lowering the vulnerability of the Western soldier to attack would be achieved in the extreme by removing her entirely from the battlefield, and in fact, new technologies may allow ground-based warfare without people. "Crewless combat" entails small, lightweight, land- and air-based combat vehicles that are controlled remotely. International law of warfare simply requires that combatants be able to distinguish combatants from non-combatants or those disabled or willing to surrender—eminently feasible with today's technologies, and even more so tomorrow.

Nanotechnology, and broader military-applied technologies, could further the West's ability to wage a "clean" war, one whose domestic cost was financial alone, not human. How would nations for whom war was bloodless—on their side—pick their battles?

Energy and the Environment: Reduced Global Interdependence

Energy and the environment are high on the public agenda, and with good reason. The energy stored in fossil fuels is finite and polluting upon release. Today the United States imports over half of its oil. U.S. policies in the Middle East—including invasion of Iraq and appeasement of Saudi Arabia—are influenced by this pragmatic reality. How would the American economy, and the global political role the nation carves out for itself, change if the United States were energetically self-sufficient?

Such self-reliance is imaginable with the right combination of technology breakthroughs and ambitious initiatives. The United States has conquered other enormous challenges through focused investment: the Manhattan Project and Apollo Project are examples. To send a person to the moon in the *Apollo,* $100 billion had to be invested. The mission in Iraq is costing the United States hundreds of billions of dollars. Clearly the country has the funds and the will to make such huge investments. Yet, in November 2004, Congress cut the budget of the U.S. National Science Foundation (NSF) by 1.9%, bringing it down to just over $5 billion. The NSF invests across basic and applied sciences, covering all fields of physical sciences—chemistry, biology, physics, all branches of engineering.

Such cuts may help to explain why we have not yet had our *Apollo*-like breakthrough in energy. Many argue for launching a $100- to $200-billion "second front" to free the United States from its dependence on oil. The fruits of such an investment are visible on the horizon. The chapter on energy presented progress in and around nanotechnology that could assist us in the necessary transition towards renewable energy as our fossil fuel extraction peaks. Large-area solar cells that were both cheap and efficient could harness power from the sun, the source of ten thousand times more energy hitting the Earth than we consume across all forms today. Solar cells would never be sufficient on their own, for their supply of energy is directly linked to the sun and we insist on consuming energy even at night and on cloudy days. Thus, technologies that store energy for later use must go hand in hand with energy-harvesting methods, and these energy-storage methods should be clean. Storage and conversion back to electrical power using technologies such as hydrogen fuel cells are thus a necessary part of an energy strategy.

To ask whether a single source could address all of our energy needs is intellectually interesting, but in reality the future of energy will look like a mosaic of generation and storage technologies—a mosaic cleaner and more sustainable than today's. Hydrogen, solar, and wind energy are attractive but their technology and economics are uncertain. Nuclear energy, "clean coal" in which carbon dioxide produced in burning coal is sequestered, and natural gas are less appealing environmentally, but better than some of what we use today, and they may form an important part of a transition phase. Changes in habits, including energy conservation, could have tremendous positive impact: as one illustration, 20% of electricity is consumed for lighting, and advanced lighting advances that replace incandescent and fluorescent lights with more efficient light-emitting diodes could reduce worldwide energy consumption by 10% and save $100 billion per year.

The direct benefits of a sustainable, clean energy strategy speak for themselves; some of their concomitant effects might get more mixed reviews. An energy self-reliant North America could, unburdened by pragmatic concerns, elect to take a purely principle-driven stance in global politics, the economic stakes of its perception in the global community lowered by its energy self-sufficiency.

An energy-sufficient North America could elect to take a purely principle-driven stance in global politics.

Nanotechnology and Choice: It's Not How Small It Is, It's How You Use It

Examples of nanotechnology's present reality and future possibility, drawn from health, the environment, and information, illustrate a point common to all technological advances. Any powerful instrument can be used for tremendous good or unfathomed ill. Words can inspire nations to embrace tolerance or hatred. Information can unite or divide. Medical understanding can be used to diagnose, cure, torture, or oppress. Wealth can increase the comfort of communion or the suffering of disparity.

How do societies determine which path they choose? In general they don't: they pick both. They nevertheless exercise some choice in the relative weighting, the proportion of good and evil done, and they are better off making the choice while technology's implements are being honed. It is incumbent upon innovators in nanotechnology to communicate the profound intellectual interest and the personally relevant potential of their work. It is imperative that nations pay attention to this information and act on it domestically and globally. And it is necessary for the people of democratic nations not to be swayed by irrational exuberance or disproportionate despondency, but instead to thrust themselves into debates over science and engineering—and our growing mastery over our natural world and physical fate—with the passion we owe to our future.

References

INTRODUCTION: Discover

p. 8 **"That night Jim and his wife ..."** H.W. Kroto, "C60—The Third Man," *Current Contents,* vol. 24, no. 36, 1993, pp. 8–9.

p. 9 **"'It was so beautiful it just had to be right.'"** H.W. Kroto, "C60—The Third Man," *Current Contents*/Engineering, Technology & Applied Sciences, vol. 24, no. 36, 1993, pp. 8–9.

p. 9 **"'During experiments aimed at ...'"** H.W. Kroto, J.R. Heath, S.C. O'Brien, R.F. Curl, R.E. Smalley, "C60: Buckminsterfullerene," *Nature,* 318, 1985, pp. 162–63.

p. 15 **"'Atoms are letters. Molecules are the words....'"** Nobel Prize interview [video] available on Nobel Prize Web site: http://nobelprize.org/chemistry/laureates/1987/

p. 18 **"'... his fundamental work in electron optics ...'"** Nobel Prize Web site: http://nobelprize.org/physics/laureates/1986/

p. 20 **"The images produced using scanning tunneling microscopes ..."** H.C. Manoharan, C.P. Lutz, D.M. Eigler, "Quantum Mirages Formed by Coherent Projection of Electronic Structure," *Nature,* vol. 403, 3 February 2000, pp. 512–15.

p. 14 **"... with breathtaking images coming out of the lab of Paul Alivisatos ..."** Liberato Manna, Delia J. Milliron, Andreas Meisel, Erik C. Scher, and A. Paul Alivisatos, "Controlled Growth of Tetrapod-Branched Inorganic Nanocrystals," *Nature Materials,* vol. 2, June 2003, pp. 382–85.

Delia J. Milliron, Steven M. Hughes, Yi Cai, Liberato Manna, Jingbo Li, Lin-Wang Wang, and A. Paul Alivisatos, "Colloidal Nanocrystal Heterostructures with Linear and Branched Topology," *Nature,* vol. 430, 8 July 2004, pp. 190–95.

General References:

M.S. Dresselhaus, G. Dresselhaus, and A. Jorio, "Unusual Properties and Structure of Carbon Nanotubes," *Annual Review of Materials Research,* 2004, vol. 34, pp. 247–78.

C.B. Murray, C.R. Kagan, M.G. Bawendi, "Synthesis and Characterization of Monodisperse Nanocrystals and Close-Packed Nanocrystal Assemblies," *Annual Review of Materials Science,* 2000, vol. 30, pp. 545–610.

George M. Whitesides and Bartosz Grzybowski, "Self-assembly at All Scales," *Science,* 29 March 2002, vol. 295, pp. 2418–21.

HEALTH

p. 31 **"North America is investing $6 billion in science and engineering research in 2005 and over $30 billion in health research."** http://www.biomedcentral.com/news/20040203/03/

1 Diagnose

p. 34 **"According to David Ahlquist ..."** http://www.fda.gov/fdac/features/2000/600_colon.html

p. 36 **"In the summer of 2004, Prof. Shuming Nie ..."** X. Gao, Y. Cui, R.M. Levenson, L.W.K. Chung, S. Nie, "In Vivo Cancer Targeting and Imaging with Semiconductor Quantum Dots," *Nature Biotechnology,* vol. 22, no. 8, August 2004, pp. 969–76.

p. 40 **"Nie has also made optical nanobarcodes, a collection of complex, multicolored, multi-quantum-dot optical labels."** Mingyong Han, Xiaohu Gao, Jack Z. Su, and Shuming Nie, "Quantum-Dot-Tagged Microbeads for Multiplexed Optical Coding of Biomolecules," *Nature Biotechnology,* vol. 19, July 2001, pp. 631–35.

p. 41 **"My group at the University of Toronto recently announced that it is possible to build beacons in the infrared that keep their brightness in blood plasma over days and weeks."** Larissa Levina, Vlad Sukhovatkin, Sergei Musikhin, Sam Cauchi, Rozalia Nisman, David P. Bazett-Jones, Edward H. Sargent, "Efficient Infrared Emitting PbS Quantum Dots Grown on DNA and Stable in Aqueous and Blood Plasma," *Advanced Materials,* 2005, DOI: 10.1002/adma.200401197.

General References:

Paul Alivisatos, "The Use of Nanocrystals in Biological Detection," *Nature Biotechnology,* vol. 22, no. 1, January 2004, pp. 47–52.

James R. Heath, Michael E. Phelps, Leroy Hood, "NanoSystems Biology," *Molecular Imaging and Biology,* vol. 5, no. 5, 2003, pp. 312–25.

Stephen R. Quake and Axel Scherer, "From Micro- to Nanofabrication with Soft Materials," *Science,* vol. 290, 24 November 2000, pp. 1536–40.

Todd Thorsen, Sebastian J. Maerkl, Stephen R. Quake, "Microfuidic Large-Scale Integration," *Science,* vol. 298, 18 October 2002, pp. 580–84.

George M. Whitesides, "The 'Right' Size in Nanobiotechnology," *Nature Biotechnology,* vol. 21, no. 10, October 2003, pp. 1161–64.

Catherine Zandonella, "The Tiny Toolkit," *Nature,* vol. 423, 1 May 2003, pp. 10–12.

2 Heal

p. 56 **"At MIT, Bob Langer, working with Professor Michael Cima, reported in 1999 a new success in controlled drug release."** J.T. Santini, Jr., M.J. Cima, and R. Langer, "A Controlled-release Microchip," *Nature,* vol. 397, 20 July 1999, pp. 335–38.

General References:

Dennis E.J.G.J. Dolmans, Dai Fukumura, and Rakesh K. Jain, "Photodynamic Therapy for Cancer," *Nature Reviews—Cancer,* vol. 3, May 2003, pp. 380–87.

Robert Langer, "Transdermal Drug Delivery: Past Progress, Current Status, and Future Prospects," *Advanced Drug Delivery Reviews*, vol. 56, 2004, pp. 557–58.

Robert Langer and David A. Tirrell, "Designing Materials for Biology and Medicine," *Nature*, vol. 428, 1 April 2004, pp. 487–92.

Marsha A. Moses, Henry Brem, and Robert Langer, "Advancing the Field of Drug Delivery: Taking Aim at Cancer," *Cancer Cell*, November 2003, vol. 4, pp. 337–41.

3 Grow

p. 62 **"In North America more than 80,000 people are awaiting organ transplantation."** Sophie Petit-Zeman, "Regenerative Medicine," *Nature Biotechnology*, vol. 19, no. 3, March 2001, pp. 201–6.

p. 62 **"Some 15% of the potential candidates for liver or heart ..."** U.A. Stock, J.P. Vacanti, "Tissue Engineering: Current State and Prospects," *Annual Review of Medicine*, 522001, pp. 443–51.

p. 66 **"At the University of Michigan, Peter Ma has shown ..."** Peter X. Ma, "Scaffolds for Tissue Fabrication," *Materials Today*, May 2004, pp. 30–40.

p. 67 **"One recent proof of this point came from George Whitesides and Donald Ingber at Harvard University."** Christopher S. Chen, Milan Mrksich, Sui Huang, George M. Whitesides, and Donald E. Ingber, "Geometric Control of Cell Life and Death," *Science*, vol. 276, 30 May 1997, pp. 1425–28.

p. 70 **"Stem cells are freshmen that have yet to specialize...."** Jochen Ringe, Christian Kaps, Gerd-Rudiger Burmester, and Michael Sittinger, "Stem Cells for Regenerative Medicine: Advances in the engineering of tissues and organs," *Naturwissenchaften*, vol. 89, 2002, pp. 338–51.

p. 73 **"Bob Langer at MIT recently showed how tissue engineering could be used to tackle the biggest problem in modern North American health."** L.E. Niklason, J. Gao, W.M. Abbott, K.K. Hirschi, S. Houser, R. Marini, and R. Langer, "Functional Arteries Grown in Vitro," *Science*, vol. 284, 16 April 1999, pp. 489–93.

p. 74 **"The same team also recently reported progress towards creating a tissue-engineered stomach."** Tomoyuki Maemura, Michael Shin, Osamu Ishii, Hidetaka Mochizuki, and Joseph P. Vacanti, "Initial Assessment of a Tissue Engineering Stomach Derived from Syngenic Donors in a Rat Model," *ASAOI Journal*, 2004, pp. 468–72.

Tomoyuki Maemura, Michael Shin, Michio Sato, Hidetaka Mochizuki, and Joseph P. Vacanti, "A Tissue-Engineered Stomach as a Replacement of the Native Stomach," *Transplantation*, vol. 76, no. 1, 15 July 2003, pp. 61–65.

p. 75 **"Subsequently, in 2004 the same team showed they could produce bone for use in reconstructive surgery."** Haru Abukawa, Michael Shin, W. Bradford Williams, Joseph P. Vacanti, Leonard B. Kaban, and Maria J. Troulis, "Reconstruction of Mandibular Defects with Autologous Tissue-Engineered Bone," *Journal of Oral and Maxillofacial Surgery*, 62, 2004, pp. 601–6.

p. 75 **"Vacanti has recently grown new large intestine...."** Tracy C. Grikscheit, Erin R. Ochoa, Anthony Ramsanahie, Eben Alsberg, David Mooney, Edward E. Whang, and Joseph P. Vacanti, "Tissue-Engineered Large Intestine Resembles Native Colon with Appropriate In Vitro Physiology and Architecture," *Annals of Surgery,* vol. 238, no. 1, July 2003, pp. 35–41.

General References:

Malcolm R. Alison, Richard Poulsom, Stuart Forbes, and Nicholas A. Wright, "An Introduction to Stem Cells," *Journal of Pathology,* vol. 197, 2002, pp. 419–23.

Anthony Atala and Chester Koh, "Tissue Engineering Applications of Therapeutic Cloning," *Annual Review of Biomedical Engineering,* vol. 6, 2004, pp. 27–40.

Paolo Bianco and Pamela Gehron Robey, "Stem Cells in Tissue Engineering," *Nature,* vol. 414, 1 November 2001, pp. 118–21.

Anne E. Bishop, Lee D.K. Buttery, and Julia M. Polak, "Embryonic Stem Cells," *Journal of Pathology,* vol. 197, 2002, pp. 424–29.

Linda G. Griffith and Gail Naughton, "Tissue Engineering—Current Challenges and Expanding Opportunities," *Science,* vol. 295, 8 February 2002, pp. 1009–14.

E. Lavik, R. Langer, "Tissue Engineering: Current State and Perspectives," *Applied Microbiology and Biotechnology,* vol. 65, 2004, pp. 1–8.

Erin R. Ochoa and Joseph P. Vacanti, "An Overview of the Pathology and Approaches to Tissue Engineering," *Annals of the New York Academy of Sciences,* 979, 2002, pp. 10–26.

David L. Stocum, "Stem Cells in Regenerative Biology and Medicine," *Wound Repair and Regeneration,* vol. 9, no. 6, November–December 2001, pp. 429–42.

Viola Vogel and Gretchen Baneyx, "The Tissue Engineering Puzzle: A Molecular Perspective," *Annual Review of Biomedical Engineering,* vol. 5, 2003, pp. 441–63.

Catherine Zandonella, "The Beat Goes On," *Nature,* vol. 421, 27 February 2003, pp. 884–86.

ENVIRONMENT

4 Energize

p. 89 **"Michael Grätzel borrowed this natural engineering strategy."** U. Bach, D. Lupo, P. Comte, J.E. Moser, F. Weissortel, J. Salbeck, H. Spreitzer, and M. Grätzel, "Solid-state Dye-Sensitized Mesoporous TiO2 Solar Cells with High Photon-to-Electron Conversion Efficiencies," *Nature,* vol. 395, 8 October 1998, pp. 583–85.

p. 92 **"In 2005, my group at the University of Toronto showed that we could turn infrared power into electricity in a paintable solar cell."** S.A. McDonald, P.W. Cyr, L. Levina, and E.H. Sargent, "Solution-Processed PbS Quantum Dot Infrared Photodetectors and Photovoltaics," *Nature Materials,* vol. 4, 9 January 2005, pp. 138–42.

General References:

Paul M. Grant, "Hydrogen Lifts Off—with a Heavy Load," *Nature,* vol. 424, 10 July 2003, pp. 129–30.

Michael Grätzel, "Photoelectrochemical Cells," *Nature*, vol. 414, 15 November 2001, pp. 338–44.

Alan J. Heeger, "Semiconducting and Metallic Polymers: The Fourth Generation of Polymeric Materials" (Nobel Lecture), *Angewandte Chemie International Edition*, vol. 40, 2001, pp. 2591–611.

Wendy U. Huynh, Janke J. Dittmer, and A. Paul Alivisatos, "Hybrid Nanorod-Polymer Solar Cells," *Science*, vol. 295, 29 March 2002, pp. 2425–27.

Mark Schrope, "Which Way to Energy Utopia?" *Nature*, vol. 414, 13 December 2001, pp. 682–84.

Brian C.H. Steele, Angelika Heinzel, "Materials for Fuel-Cell Technologies," *Nature*, vol. 414, 15 November 2001, pp. 345–52.

5 Protect

p. 105 **"Seacoast Science in Carlsbad, near San Diego ..."** Jonathan Knight, "Tomorrow's World," *Nature*, vol. 426, 11 December 2003, pp. 709–11.

p. 105 **"Tim Swager and his team at MIT have taken a different approach to the sensitive measurement of specific neurotoxins."** Anthony W. Czarnik, "A Sense for Landmines," *Nature*, vol. 394, 30 July 1998, pp. 417–18.

Shi-Wei Zhang and Timothy M. Swager, "Fluorescent Detection of Chemical Warfare Agents: Functional Group Specific Ratiometric Chemosensors," *Journal of the American Chemical Society*, vol. 125, 2003, pp. 3420–21.

p. 111 **"In fact, environmental remediation may be among the greatest applications for zeolites."** U.S. EPA 2004 Nanotechnology Science to Achieve Results (STAR) Progress Review Workshop—Nanotechnology and the Environment II, August 18–20, 2004. http://es.epa.gov/ncer/publications/meetings/8-18-04/agenda.html

p. 112 **"Vicky Colvin at Rice University, in Houston, is looking into the toxicity of nanomaterials, focusing in particular on nanoparticles."** Vicki L. Colvin, "The Potential Environmental Impact of Engineered Nanomaterials," *Nature Biotechnology*, vol. 21, No. 10, October 2003, pp. 1166–70.

p. 112 **"She recently found that the toxicity of buckyballs depends entirely on the chemical state of the nanoparticles' surfaces."** C.M. Sayes, J.D. Fortner, W. Guo, D. Lyon, A.M. Boyd, K.D. Ausman, Y.J. Tao, B. Sitharaman, L.J. Wilson, J.B. Hughes, J.L. West, and V.L. Colvin, "The Differential Cytotoxicity of Water-soluble Fullerenes," *Nano Letters*, vol. 4, no. 10, 2004, pp. 1881–87.

p. 113 **"The first verified studies of single-walled carbon nanotube toxicity appeared in 2003."** D.B. Warheit, B.R. Laurence, K.L. Reed, D.H. Roach, G.A.M. Reynolds, and T.R. Webb, "Comparative Pulmonary Toxicity Assessment of Single-Wall Carbon Nanotubes in Rats," *Toxicological Sciences*, vol. 77, no. 1, 2004, pp. 117–25.

General References:

A. Corma, "From Microporous to Mesoporous Molecular Sieve Materials and Their Use in Catalysis," *Chemistry Reviews*, vol. 97, 1997, pp. 2373–419.

S.O. Obare and G.J. Meyer, "Nanostructured Materials for Environmental Remediation of Organic Contaminants in Water," *Journal of Environmental Science and Health—Part A—Toxic/Hazardous Substances & Environmental Engineering,* vol. A39, no. 10, 2004, pp. 2549–82.

J.D. Sherman, "Synthetic Zeolites and Other Microporous Oxide Molecular Sieves," Proceedings of the National Academy of Sciences, vol. 96, 1999, pp. 3471–78.

6 Emulate

p. 117 **"Spider silk is another example of Nature's design...."** Zhengzhong Shao and Fritz Vollrath, "Surprising Strength of Silkworm Silk," *Nature,* vol. 418, 15 August 2002, p. 741.

p. 118 **"In 1996, Belcher examined how Nature builds shell of the abalone...."** A.M. Belcher, X.H. Wu, R.J. Christensen, P.K. Hansma, G.D. Stucky, and D.E. Morse, "Control of Crystal Phase Switching and Orientation by Soluble Mollusc-Shell Proteins," *Nature,* vol. 381, 2 May 1996, pp. 56–58.

p. 119 **"Belcher looked further into why the shells are so strong."** Bettye L. Smith, Tilman E. Schaffer, Mario Viani, James B. Thompson, Neil A. Frederick, Johannes Kindt, Angela Belcher, Galen D. Stucky, Daniel E. Morse, and Paul K. Hansma, "Molecular Mechanistic Origin of the Toughness of Natural Adhesives, Fibres and Composites," *Nature,* vol. 399, 24 June 1999, pp. 761–63.

p. 120 **"Belcher chose an objective for her game of evolution ..."** Sandra R. Whaley, D.S. English, Evelyn L. Hu, Paul F. Barbara, and Angela M. Belcher, "Selection of Peptides with Semiconductor Binding Specificity for Directed Nanocrystal Assembly," *Nature,* vol. 405, 8 June 2000, pp. 665–68.

p. 121 **"Instead of waiting for Nature's occasional mutations, she would search systematically through billions of protein options in a single experiment."** Nadrian C. Seeman and Angela M. Belcher, "Emulating Biology: Building nanostructures from the bottom up," Proceedings of the National Academy of Sciences, vol. 99, suppl. 2, 30 April 2002, pp. 6451–55.

p. 122 **"Belcher obtained a library of viruses ..."** Seung-Wuk Lee, Chuanbin Mao, Christine E. Flynn, and Angela M. Belcher, "Ordering of Quantum Dots Using Genetically Engineered Viruses," *Science,* vol. 296, 3 May 2002, pp. 892–95.

p. 123 **"This technique, used by Belcher in the late 1990s to pan for proteins with specific properties ..."** Mehmet Sarikaya, Candan Temerler, Alex K.Y. Jen, Klaus Schulten, Francois Baneyx, "Molecular Biomimetics: Nanotechnology through Biology," *Nature Materials,* vol. 2, September 2003, pp. 577–85.

p. 127 **"Yale Goldman and colleagues at the University of Pennsylvania recently sought to answer the question by watching how a molecular motor walked along a track."** Ahmet Yildiz, Joseph N. Forkey, Sean A. McKinney, Taekjip Ha, Yale E. Goldman, and Paul R. Selvin, "Myosin V Walks Hand-Over-Hand: Single Fluorophore Imaging with 1.5-nm Localization," *Science,* vol. 300, 27 June 2003, pp. 2061–65.

p. 127 **"Since Jeff Gelles at Brandeis University, in Massachusetts, had previously eliminated the waddle hypothesis ..."** Wei Hua, Johnson Chung, and Jeff Gelles, "Distinguishing Inchworm and Hand-Over-Hand Processive Kinesin Movement by Neck Rotation Measurements," *Science,* vol. 295, 1 February 2002, pp. 844–48.

p. 128 **"Ross Kelly at Boston College recently made progress towards molecular motors of our own design."** T. Ross Kelly, Harshani De Silva, and Richard A. Silva, "Unidirectional Rotary Motion in a Molecular System," *Nature,* vol. 401, 9 September 1999, pp. 150–52.

p. 129 **"Bernard Yurke at Bell Laboratories in New Jersey, with collaborators at Oxford University, showed that their engines implemented continuous operation—repeated motion."** Bernard Yurke, Andrew J. Turberfield, Allen P. Mills, Jr., Friedrich C. Simmel, and Jennifer L. Neumann, "A DNA-Fuelled Molecular Machine Made of DNA," *Nature,* vol. 406, 10 August 2000, pp. 605–8.

General References:

Vincenzo Balzani, Alberto Credi, Francisco M. Raymo, and J. Fraser Stoddart, "Artificial Molecular Machines," *Angewandte Chemie International Edition,* vol. 39, 2000, pp. 3348–91.

Jose V. Hernandez, Euan R. Kay, and David A. Leigh, "A Reversible Synthetic Rotary Molecular Motor," *Science,* vol. 306, 26 November 2004, pp. 1532–37.

Nagatoshi Koumura, Robert W.J. Zijlstra, Richard A. van Delden, Nobuyuki Harada, and Ben L. Feringa, "Light-Driven Monodirectional Molecular Rotor," *Nature,* vol. 401, 9 September 1999, pp. 152–55.

Chuanbin Mao, Daniel J. Solis, Brian D. Reiss, Stephen T. Kottmann, Rozamond Y. Sweeney, Andrew Hayhurst, George Georgiou, Brent Iverson, and Angela M. Belcher, "Virus-Based Toolkit for the Directed Synthesis of Magnetic and Semiconducting Nanowires," *Science,* vol. 303, 9 January 2004, pp. 213–17.

Jeffrey M. Perkel, "Investigating Molecular Motors Step by Step," *The Scientist,* vol. 15, 2004, pp. 19–23.

Manfred Schliwa and Gunther Woehlke, "Molecular Motors," *Nature,* vol. 422, 17 April 2003, pp. 759–65.

Nadrian C. Seeman, "DNA in a Material World," *Nature,* vol. 421, 23 January 2003, pp. 427–31.

William B. Sherman and Nadrian C. Seeman, "A Precisely Controlled DNA Biped Walking Device," *Nano Letters,* vol. 4, no. 7, 2004, pp. 1203–7.

INFORMATION

7 Compute

pp. 138–39 **"Gordon Moore, founder of chip-maker Intel, showed legendary prescience in his 1965 article titled 'Cramming More Components onto Integrated Circuits.'"** Gordon Moore, "Cramming More Components onto Integrated Circuits," *Electronics,* vol. 38, no. 8, 1965, pp. 114–17.

p. 141 **"As Rebecca Henderson of MIT's Sloan School of Management pointed out in her 1995 work ..."** Rebecca Henderson, "Of Life Cycles Real and

Imaginary: The Unexpectedly Long Old Age of Optical Lithography," *Research Policy,* vol. 24, 1995, pp. 631–43.

p. 144 **"The vertical relationship between gate and channel is critical too."** Max Schultz, "The End of the Road for Silicon?" *Nature,* vol. 399, 24 June 1999, pp. 729–30.

Seth Lloyd, "Ultimate Physical Limits to Computation," *Nature,* vol. 406, 31 August 2000, pp. 1047–54.

p. 146 **"Bulked-up insulating materials could further extend the persistent enforcement of Moore's Law."** Clemens J. Forst, Christopher R. Ashman, Karlheinz Schwarz, and Peter E. Blochl, "The Interface between Silicon and a High-k Oxide," *Nature,* vol. 427, 1 January 2004, pp. 53–56.

p. 148 **"Cees Dekker and colleagues in The Netherlands demonstrated in 1997 a transistor in which the channel was formed by one molecule alone."** Sander J. Tans, Michel H. Devoret, Hongjie Dai, Andreas Thess, Richard E. Smalley, L. J. Geerligs, and Cees Dekker, "Individual Single-Wall Carbon Nanotubes as Quantum Wires," *Nature,* vol. 386, 3 April 1997, pp. 474–77.

p. 152 **"This strategy would need to rely on electrically programmable molecular diodes ..."** Charles P. Collier, Gunter Mattersteig, Eric W. Wong, Yi Luo, Kristen Beverly, Jose Sampaio, Francisco M. Raymo, J. Fraser Stoddart, and James R. Heath, "A [2]Catenane-Based Solid State Electronically Reconfigurable Switch," *Science,* vol. 289, 18 August 2000, pp. 1172–75.

C.P. Collier, E.W. Wong, M. Belohradsky, F. M. Raymo, J.F. Stoddart, P.J. Kuekes, R.S. Williams, and J.R. Heath, "Electronically Configurable Molecular-Based Logic Gates," *Science,* vol. 285, 16 July 1999, pp. 391–95.

p. 152 **"Critics have suggested that voltages intended to turn switchable molecules on and off may have changed the properties of the metal electrodes themselves."** Robert F. Service, "Molecular Electronics: Next-generation Technology Hits an Early Midlife Crisis," *Science,* vol. 302, 24 October 2003, pp. 556–58.

p. 153 **"Stan Williams of HP labs, Hewlett-Packard's central research organization, has focused on building computers not using three-terminal transistors as in today's machines, but rather using Aviram and Ratner's simpler molecular diodes."** James R. Heath, Philip J. Kuekes, Gregory S. Snider, R. Stanley Williams, "A Defect-Tolerant Computer Architecture: Opportunities for Nanotechnology," *Science,* vol. 280, 12 June 1998, pp. 1716–21.

p. 153 **"Researchers at the California NanoSystems Institute (CNSI) have shown one way to deal with the wiring problem...."** Nicholas A. Melosh, Akram Boukai, Frederic Diana, Brian Gerardot, Antonio Badlato, Pierre M. Petroff, and James R. Heath, "Ultrahigh-Density Nanowire Lattices and Circuits," *Science,* vol. 300, 4 April 2003, pp. 112–15.

General References:

Philip Ball, "Chemistry Meets Computing," *Nature,* vol. 406, 13 July 2000, pp. 118–20.

James R. Heath and Mark A. Ratner, "Molecular Electronics," *Physics Today*, May 2003, pp. 43–49.

Zhen Yao, Henk W. Ch. Postma, Leon Balents, and Cees Dekker, "Carbon Nanotube Intramolecular Junctions," *Nature*, vol. 402, 18 November 1999, pp. 273–76.

Takashi Ito and Shinji Ikazaki, "Pushing the Limits of Lithography," *Nature*, vol. 406, 31 August 2000, pp. 1027–31.

C. Joachim, J.K. Gimzewski, and A. Aviram, "Electronics Using Hybrid-Molecular and Mono-Molecular Devices," *Nature*, vol. 408, 30 November 2000, pp. 541–48.

Mark Lundstrom, "Moore's Law Forever?" *Science*, vol. 299, 10 January 2003, pp. 210–11.

Dennis Normile, "The End—Not Here Yet, But Coming Soon," *Science*, vol. 293, 3 August 2001, p. 787.

Paul S. Peercy, "The Drive to Miniaturization," *Nature*, vol. 406, 31 August 2000, pp. 1023–26.

8 Interact

p. 162 **"In 1989, Richard Friend of Cambridge University found that polymers, when energized with electricity, could be made to produce light."** J.H. Burroughes, D.D.C. Bradley, A.R. Brown, R.N. Marks, K. Mackay, R.H. Friend, P.L. Burns, and A.B. Holmes, "Light-Emitting Diodes Based on Conjugated Polymers," *Nature*, vol. 347, 11 October 1990, pp. 539–41.

p. 162 **"At the heart of the device was a semiconducting polymer layer ..."** P.L. Burn, A.B. Holmes, A. Kraft, D.D.C. Bradley, A.R. Brown, R.H. Friend, R.W. Gymer, "Chemical Tuning of Electroluminescent Copolymers to Improve Emission Efficiencies and Allow Patterning," *Nature*, vol. 356, 5 March 1992, pp. 47–49.

p. 163 **"Printable light-emitting polymers have been made to cover the range ..."** H. Sirringhaus, T. Kawase, R.H. Friend, T. Shimoda, M. Inbasekaran, W. Wu, and E.P. Woo, "High-Resolution Inkjet Printing of All-Polymer Transistor Circuits," *Science*, vol. 290, 15 December 2000, pp. 2123–26.

p. 163 **"Another way to tune color unites long-chain polymers with smaller organic molecules ..."** Mounir Halim, Jonathan N.G. Pillow, Ifor D.W. Samuel, and Paul L. Burn, "Conjugated Dendrimers for Light-Emitting Diodes: Effect of Generation," *Advanced Materials*, vol. 11, no. 5, 1999, pp. 371–74.

p. 163 **"Vladimir Bulovic at MIT built a device with an active layer made up of only a single layer of quantum dots, but clad by organic semiconductors."** Seth Coe, Wing-Keung Woo, Moungi Bawendi, and Vladimir Bulovic, "Electroluminescence from Single Monolayers of Nanocrystals in Molecular Organic Devices,' *Nature*, vol. 420, December 2002, pp. 800–3.

p. 166 **"Mark Humayun and colleagues at USC therefore set about building a retinal prosthesis, a system that would include a camera to acquire images and**

a system to stimulate, using electrical pulses, the neurons of the retina." Mark S. Humayun, James D. Weiland, Gildo Y. Fujii, Robert Greenberg, Richard Williamson, Jim Little, Brian Mech, Valerie Cimmarusti, Gretchen Van Boemel, Gislin Dagnelie, and Eugene de Juan, Jr., "Visual Perception in a Blind Subject with a Chronic Microelectronic Retinal Prosthesis," *Vision Research*, vol. 43, 2003, pp. 2573–81.

General References:

Ananth Dodabalapur, "Betting on Organic CMOS," *Materials Today*, September 2004, p. 56.

R.H. Friend, R.W. Gymer, A. Holmes, J.H. Burroughes, R.N. Marks, C. Taliani, D.D.C. Bradley, D.A. Dos Santos, J.L. Bredas, M. Logdlund, W.R. Salaneck, "Electroluminescence in Conjugated Polymers," *Nature*, vol. 397, 14 January 1999, pp. 121–29.

Karl Ziemelis, "Putting It on Plastic," *Nature*, vol. 393, 18 June 1998, pp. 619–20.

9 Convey

p. 179 **"Partha Mitra and Jason Stark of Bell Laboratories explored the information theoretic limits of optical fiber."** Partha P. Mitra and Jason B. Stark, "Nonlinear Limits to the Information Capacity of Optical Fibre Communications," *Nature*, vol. 411, 28 June 2001, pp. 1027–30.

p. 184 **"... and a team of experimentalists set about turning Kuzyk's dream into reality."** Q. Chen, L. Kuang, Z.Y. Wang, and E.H. Sargent, "Crosslinked C60-Polymer Breaches the Quantum Gap," *Nano Letters*, vol. 4, no. 9, 2004, pp. 1673–75.

p. 186 **"Yoshio Yamamoto of Stanford University in California has built extremely reflective mirrors and has paired them up such that light is trapped between them...."** G.S. Solomon, M. Pelton, and Y. Yamamoto, "Modification of Spontaneous Emission of a Single Quantum Dot," *Physica Status Solidi*, vol. 178, 2000, pp. 341–44.

p. 187 **"Kerry Vahala and his team at Caltech in Pasadena, California, have built not a cage for light, but a racetrack."** D.K. Armani, T.J. Kippenberg, S.M. Spillane, and K.J. Vahala, "Ultra-High-Q Toroid Microcavity on a Chip," *Nature*, vol. 421, 27 February 2003, pp. 925–28.

p. 191 **"A recent triumph for the top-down camp came from the efforts of Hank Smith and collaborators at MIT."** Minghao Qi, Elefterios Lidorikis, Peter T. Rakich, Steven G. Johnson, J.D. Joannopoulos, Erich P. Ippen, and Henry I. Smith, "A Three-Dimensional Optical Photonic Crystal with Designed Point Defects," *Nature*, vol. 429, 3 June 2004, pp. 538–42.

p. 192 **"Among the bottom-ups, Sajeev John, Geff Odin, Henry van Driel, and their collaborators in Toronto and Spain had a major success in 2001."** Alvaro Blanco, Emmanuel Chomski, Serguei Grabtchak, Marta Ibisate, Sajeev John, Stephen W. Leonard, Cefe Lopez, Francisco Meseguer, Herman Miguez, Jessica P. Mondia, Geoffrey A. Ozin, Ovidiu Toader, and Henry M. van Driel, "Large-Scale Synthesis of a Silicon Photonic Crystal with a Complete Three-

Dimensional Bandgap Near 1.5 Micrometers," *Nature,* vol. 405, 25 May 2000, pp. 437–40.

p. 194 **"Two years later, a group at Intel, using the techniques already available to produce Intel's Pentium processors, showed that they could create an optical modulator—a device that puts information on to light from a laser."** Ansheng Liu, Richard Jones, Ling Liao, Dean Samura-Rubio, Doron Rubin, Oded Cohen, Remus Nicolaescu, and Mario Paniccia, "A High-Speed Silicon Optical Modulator Based on a Metal-Oxide-Semiconductor Capacitor," *Nature,* vol. 427, 12 February 2004, pp. 615–18.

p. 194 **"In 2000, a group at the University of Trento in Italy observed optical gain—the key ingredient in making a laser—from silicon nanocrystals."** L. Pavesi, L. Dal Negro, C. Mazzoleni, G. Franzo, and F. Priele, "Optical Gain in Silicon Nanocrystals," *Nature,* vol. 408, 23 November 2000, pp. 440–44.

p. 195 **"Collectively, the teams led by Uri Banin at Hebrew University of Jerusalem; my group at the University of Toronto; and the collaborators Vladimir Bulovic and Moungi Bawendi at MIT have reported significant progress since 2001 along this path."** Nir Tessler, Vlad Medvedev, Miri Kazes, ShiHai Kan, and Uri Banin, "Efficient Near-Infrared Polymer Nanocrystal Light-Emitting Diodes," *Science,* vol. 295, 22 February 2002, pp. 1506–8.

L. Bakueva, S. Musikhin, M.A. Hines, T.W.F. Chang, M. Tzolov, G.D. Scholes, E.H. Sargent, "Size-Tunable Infrared (1000–1600 nm) Electroluminescence from PbS Quantum-Dot Nanocrystals in a Semiconducting Polymer," *Applied Physics Letters,* vol. 82, no. 17, 2003, pp. 2895–97.

J.S. Steckel, S. Coe-Sullivan, V. Bulovic, M.G. Bawendi, "1.3 mu to 1.55 mu Tunable Electroluminescence from PbSe Quantum Dots Embedded within an Organic Device," *Advanced Materials,* vol. 15, no. 21, 2003, pp. 1862–66.

Tung-Wah Frederick Chang, Sergei Musikhin, Ludmila Bakueva, Larissa Levina, Margaret A. Hines, Paul W. Cyr, and Edward H. Sargent, "Efficient Excitation Transfer from Polymer to Nanocrystals," *Applied Physics Letters,* vol, 84, no. 21, 2004, pp. 4295–97.

Steven A. McDonald, Gerasimos Konstantatos, Shiguo Zhang, Paul W. Cyr, Ethan J.D. Klem, Larissa Levina, and Edward H. Sargent, "Solution-Processed PbS Quantum Dot Infrared Photodetectors and Photovoltaics," *Nature Materials,* vol. 4, no. 2, 2005, 138–42.

Edward H. Sargent, "Infrared Quantum Dots," *Advanced Materials,* vol. 17, March 8, 2005, pp. 515–22.

Brian L. Wehrenberg and Philippe Guyot-Sionnest, "Electron and Hole Injection in PbSe Quantum Dot Films," *Journal of the American Chemical Society,* vol. 125, 2003, pp. 7806–7.

General References:

Peihong Zhang, Vincent H. Crespi, Eric Chang, Steven G. Louie, and Marvin L. Cohen, "Computational Design of Direct-Bandgap Semiconductors that Lattice-Match to Silicon," *Nature,* vol. 409, 4 January 2001, pp. 69–71.

EPILOGUE: Humanize

General References:

Jurgen Altmann and Mark Gubrud, "Anticipating Military Nanotechnology," *IEEE Technology and Society Magazine*, Winter 2004, pp. 33–40.

M.E. Gorman, J.F. Groves, and R.K. Catalano, "Society Dimensions of Nanotechnology," *IEEE Technology and Society Magazine*, Winter 2004, pp. 55–62.

Robert F. Service, "Nanotechnology Grows Up," *Science*, vol. 304, 18 June 2004, pp. 1732–34

Illustration Credits

Page

21 H.C. Manoharan, C.P. Lutz, D.M. Eigler, et al, "Quantum Mirages Formed by Coherent Projection of Electronic Structure," *Nature,* vol. 403, pp. 512–15. Used with permission.

39 X. Gao, Y. Cui, R.M. Levenson, L.W.K. Chung, S. Nie, et al, "In Vivo Cancer Targeting and Imaging with Semiconductor Quantum Dots," *Nature,* vol. 22, pp. 969–76. Used with permission.

57 John T. Santini Jr., Michael J. Cima, Robert Langer, et al., "A Controlled-Release Microchip," *Nature,* vol. 397, pp. 335–38. Used with permission.

68 Reprinted with permission from Gabriel A. Silva, Catherine Czeister, Krista L. Niece, Elia Beniash, Daniel A. Harrington, John A. Kessler, and Samuel I. Stupp, "Selective Differentiation of Neural Progenitor Cells by High-Epitope Density Nanofibers," *Science,* vol. 303, pp. 1352–55 (2004), published online 22 January 2004 (10.1126/science.1093783). Copyright 2004, AAAS.

85 Reprinted with permission of *Technology Review.*

123 Reprinted with permission from Chuanbin Mao, Daniel J. Solis, Brian D. Reiss, Stephen T. Kottmann, Rozamond Y. Sweendy, Andrew Hayhurst, George Georgiou, Brent Iverson, and Angela M. Belcher, "Virus-Based Toolkit for the Directed Synthesis of Magnetic and Semiconducting Nanowires," *Science,* vol. 303, pp. 213–17 (2004). Copyright 2004, AAAS.

130 Bernard Yurke, Andrew J. Turberfield, Allen P. Mills Jr., Friedrich C. Simmel, Jennifer L. Neumann, et al., "A DNA-Fuelled Molecular Machine Made of DNA," *Nature,* vol. 406, pp. 605–8. Used with permission.

144 Paul S. Peercy, et al., "The Drive to Miniaturization," *Nature,* vol. 406, pp. 1023–26. Used with permission.

154 Reprinted with permission from Nicholas A. Melosh, Akram Boukai, Frederic Diana, Brian Gerardot, Antonio Badolato, Pierre M. Petroff, and

James R. Heath, "Ultrahigh-Density Nanowire Lattices and Circuits," *Science*, vol. 300, pp. 112–15 (2003), published online 13 March 2003 (10.1126/science.1081940). Copyright 2003, AAAS.

166 Reprinted from *Vision Research*, vol. 43, Mark S. Humayun, James D. Weiland, Gildo Y. Fujii, Robert Greenberg, Richard Williamson, Jim Little, Brian Mech, Valerie Cimmarusti, Gretchen Van Boemel, Gislin Dagnelie, Eugene de Juan Jr., "Visual Perception in a Blind Subject with a Chronic Microelectronic Retinal Prosthesis," pp. 2573–81, 2003, with permission from Elsevier.

186, 187 Kerry J. Vahala, et al., "Optical Microcavities," *Nature*, vol. 424, pp. 839–46. Used with permission.

193 Alvaro Blanco, Emmanuel Chomski, Serguei Grabtchak, Marta Ibisate, et al., "Large-Scale Synthesis of a Silicon Photonic Crystal with a Complete Three-Dimensional Bandgap Near 1.5 Micrometers," *Nature*, vol. 405, p. 437. Used with permission.

Acknowledgments

David Davidar, Susan Folkins, Sandra Tooze, Allyson Latta, Colleen Clarke, and the entire team at Penguin Canada; and Jofie Ferrari-Adler and John Oakes at Thunder's Mouth/Avalon; and Westwood Creative Artists' Bruce Westwood, Ashton Westwood, and Nicole Winstanley. It was a thrill to work with you—thank you for believing in the book and in me.

Shana Kelley, Guy Weichenberg, Jane Freeman, Tom Kudrle, Jen Howard, Andrew Grenville, Adrienne Leahey, Nikki Barrett, and John, Janice, and Laurie Sargent were generous with feedback and encouragement on the manuscript.

Thanks to my colleagues at the University of Toronto and my collaborators at the Massachusetts Institute of Technology: you have welcomed me generously into an exciting area of scholarship.

Index